Build your own house

Build your own house

Stuart Martin

STANLEY PAUL | LONDON

STANLEY PAUL & CO LTD
3 Fitzroy Square, London W1

AN IMPRINT OF THE HUTCHINSON GROUP

London Melbourne Sydney Auckland
Wellington Johannesburg Cape Town
and agencies throughout the world

First published March 1960
Second impression March 1961
Third impression October 1964
Fourth Revised Edition January 1969
Fifth Revised Edition December 1972
Sixth impression September 1973

Printed in Great Britain by litho on smooth wove paper
by Anchor Press, and bound by Wm. Brendon,
both of Tiptree, Essex
ISBN 0 09 110170 0

Contents

Author's Note

While every effort has been made to ensure that the facts contained in this book are accurate, the author is unable to guarantee that the details furnished can be relied on in all circumstances.

Illustrations

Drainage detail with Y junction
Drainage details
Drainage detail, showing gully and manhole
Steps with brick-on-edge risers
Fuel-bunker under construction
Corner set up for fuel-bunker

FIGURES

Preface to fifth revised edition

The need for a fifth revised edition of Build Your Own House has been brought about mainly by the introduction of decimalisation. Consequently, all the prices quoted have been modified to the new currency and they have also been increased to suit the present upward trend. At the same time a few minor alterations have been made to the general text.

Since the publication of the last edition another new feature, Metrication, has been introduced into the building and construction industry. However, this process is not yet complete and, for the time being, it has been decided to allow the Imperial units quoted in the book to remain unaltered.

On the other hand the reader should certainly be aware that this change is taking place, and anyone embarking on a building project such as that described here will find that metric sizes and quantities are already being used for many of the materials and items that need to be purchased.

Although it has been decided to retain Imperial units in the present edition the following data should give the reader sufficient basic information to make the necessary comparison and adjustment between the two sets of measurements.

The metric rule is, of course, graduated in millimetres, while the Imperial rule measures feet and inches.

It can be seen that 300 mm. on a metric rule is approximately equal to 1 ft. on the Imperial rule (305 mm. is a more exact figure). Similarly, 200 mm., 100 mm., 50 mm. and 25 mm. are all approximately equal to 8 in., 4 in., 2 in. and 1 in. respectively. These five metric dimensions are being used as basic sizes for co-ordinating the dimensions of components used within the industry and they are known as preferred dimensions.

With doors, frames, windows and other building components dimensionally co-ordinated in this way the need for non-standard units requiring to be site trimmed will be eliminated.

Where cubic measures are concerned, large volumes such as sand, gravel, etc., will be measured in cubic metres. This metric unit of measurement shows as an increase in size when it is compared with the cubic volumetric unit of Imperial measurement, the cubic yard. A cubic metre contains 35·31 cu. ft. compared with the 27 cu. ft. of the cubic yard. Consequently, a suitable allowance will have to be made when ordering materials supplied in metric quantities.

In the same way there is an increase in the size of the unit of square measurement when going metric, a square metre containing 10·76 sq. ft. compared with the 9 sq. ft. of the Imperial square yard.

The diagram on the opposite page should help to clarify some of the points and help the reader to make the necessary conversions as he comes across them.

Softwoods are already being imported into this country in a new range of sizes in metric measurements. The section sizes are given in millimetres and the lengths in metres. These new sizes are slightly under the corresponding Imperial sizes because, as previously explained, the metric equivalents of the Imperial sizes have been 'rounded off' to suit a metric module.

The table below illustrates this:

Section sizes

Imperial measure (inches)	1×3	2×4	3×6	4×12
Metric measure (mm)	25×75	50×100	75×150	100×300

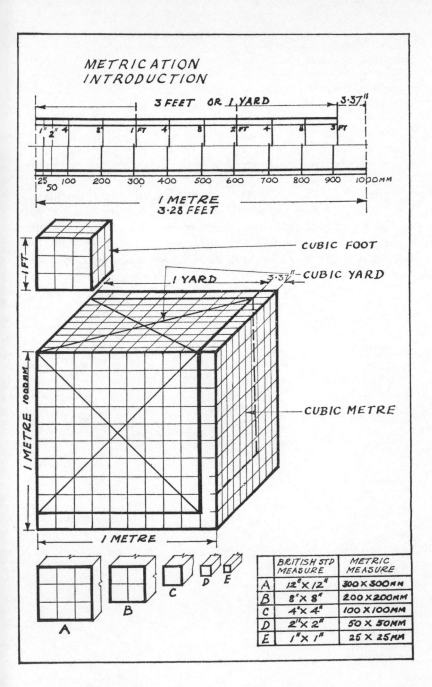

METRICATION
INTRODUCTION

3 FEET OR 1 YARD   3·37"

1" 2" 4"   8"   1 FT   4"   8"   2 FT   4"   8"   3 FT

25 100 200 300 400 500 600 700 800 900 1000MM
50

1 METRE
3·28 FEET

CUBIC FOOT

1 FT

1 YARD   3·37"—CUBIC YARD

1 METRE 100DAM

CUBIC METRE

1 METRE

	BRITISH STD MEASURE	METRIC MEASURE
A	12"×12"	300 × 300MM
B	8"× 8"	200 ×200MM
C	4"× 4"	100 ×100MM
D	2"× 2"	50 × 50MM
E	1"× 1"	25 × 25MM

Lengths

Imperial measure (feet)	8	9	10	15	20
Metric measure (metres)	2·40	2·70	3·00	4·50	6·00

Finally, to show how the new system of measurement will eventually simplify the whole approach to planning and building when it is completely in operation, a portion of a 300 mm. planning grid is shown in the following diagram. And of course, the real secret is to 'think metric' instead of always trying to make direct conversions.

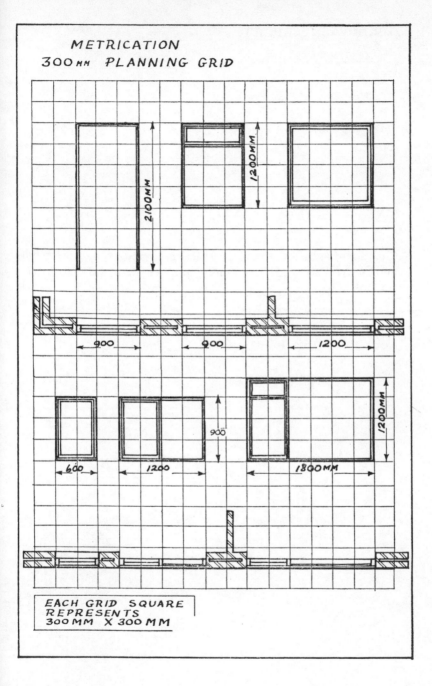

METRICATION
300 MM PLANNING GRID

2100MM
1200MM
900
900
1200
600
1200
1800MM
1200MM

EACH GRID SQUARE
REPRESENTS
300 MM X 300 MM

Acknowledgements

Grateful thanks are due to the various organizations, companies and individuals who have kindly supplied data and information for this book.

Introduction

Since building a pair of houses by the combination of hard work and the employment of direct labour, the author has been asked many questions by others engaged on, or contemplating doing, the same.

By working for oneself, or by employing direct labour, a house or bungalow can be built within the mortgage allowed on the value of the property. The mortgage may come from Building Society, Insurance Company or Local Authority. If a small dwelling has been built within this financial boundary no deposit will have been paid, as the deposit represents the difference between the mortgage allowed and the purchase price. The purchase price of a building includes the builder's profit, and by being, virtually, your own builder, the necessity for a large deposit is eliminated.

However, the person building this way must ensure that he has the means to enable him to get to that point where the first draw on the mortgage can be claimed. The largest items of expense in the initial stages are the cost of land and bricks. It may be that the land can be purchased from the Local Authority, and if the mortgage is obtained from them also they may deduct the cost of the land from the mortgage at the outset. With an item like a brick bill it may be possible to obtain credit until the first draw is obtained, or a loan from a private source, which again can be paid when the draws on the mortgage take place.

Conditions for mortgages vary according to the source from which they are obtained, and the person who contemplates building a house should, obviously, make the fullest inquiries before deciding on a course of action.

It is not proposed to deal with the financial side of the question other than the foregoing brief remarks, as there are several books published already which deal with house purchase.

Neither does the author set out to replace the architect or builder. This book is intended for those who, spurred on by desperate living conditions and with limited money available, feel that building their own house by direct, or their own, labour is the solution to their problem.

To most people, building terms are unfamiliar and the step-by-step construction of a modern house or bungalow unknown. This book sets out the stages of construction, and, if the reader is building for himself, it will help him decide which jobs he can tackle and which jobs to sub-contract. Even the person who employs a builder throughout should find these pages helpful. Having read them he should be able to discuss with the builder all the points which arise during construction on a more equal footing.

One

The site and the plans

Having decided that you are going to build a house the first step is to secure a site. This can be done either by private treaty or through the Local Authority.

It is obviously a good idea to inspect as many sites as possible and then compare their merits. A badly chosen site, apart from any aspect considerations, can involve the owner in lots of additional building expenses.

In choosing the situation of the site the main consideration will be convenience for reaching work, shops, bus services and schools for the children. In the latter instance it should be kept in mind that children often have to change schools as they advance in their education.

An item that is often overlooked is the future development of the area. New trunk roads, large council estates or new factories can soon change the appearance of a district in a few years, and the local council should be consulted as to future intentions.

Having satisfied yourself on the foregoing points, and having chosen a district, what are the next points to consider, assuming that you are lucky enough to have the choice of several sites?

First of all, is the site suitable for a house or a bungalow, remembering that the latter will generally need a wider frontage?

On any site ensure that there is room for a garage, even if you do not build it at the same time that you build your house. Even if you do not possess a car, a garage is a worth-while consideration as it is an added selling point if circumstances ever force you to leave the district. In any case, it is unlikely that your plans would receive approval under the Planning Acts unless they showed that there was the necessary space for a garage, together with a safe and suitable access from the road.

Another important consideration to be borne in mind is whether the site is fully developed. If it is, that is with roads made up, water, gas and electricity laid on and sewers in existence, the initial cost of the site may be more, but there would be no future charges. If the site is not fully developed and roads and sewers are destined for a later date, each householder would have to pay a proportion of the cost of making the road according to frontage, and the temporary lack of sewers would mean alternative drainage arrangements.

In rural districts sewerage is taken into a septic tank, and storm water into soakaways.

There may be an individual septic tank required on each site or there may be a communal one serving several sites. In either case the pros and cons of the cost should be gone into before any decision is made to build. As far as soakaways are concerned the cost is not excessive. These consist of pits filled with clinker and hardcore into which the storm-water drains are led.

Many people have a preference for a corner site, and whilst this is ideal in some respects the cost of the extra front walling and possibly the road charges may make you wish that you had been less ambitious. If the site should be sloping as well, the wall foundations would have to be stepped, and this will add further cost. The digging of such foundations are more complicated than for a straightforward wall on a level site.

The slope of the site should be considered anyway. When foundations are brought up to damp-course level, a height of 6 in. above ground at the front or back may mean a drop of several feet at the opposite side. If solid floors are proposed this means that the brickwork will have to be filled in with hardcore before the site concrete is laid, and this all adds to the cost. It also means that there are far more bricks than usual in the foundations. If there is to be no building-up round the house on the high side, these bricks will have to be facing bricks. These facing bricks are more expensive than the commons used for inside walling.

The alternative is to use facing bricks for several courses below the damp course and then to change to commons for both inside and outside skins, and banking-up all round with earth so that no common bricks are showing. This alternative is only slightly less expensive and means a lot of hard work in the banking-up.

Scaffolding is a point to consider as well. If there is a considerable slope on the site, scaffolding will be required earlier than normal, with the attendant trouble and cost.

Another point which may adversely affect the cost is the sewer position. If the sewers are on the plot, and the run to reach them is short, the excavation may be done at a reasonable cost. On the other hand, you may have to cross a road to make your connection, and if this is the case part of the work will have to be done by the Local Authority, and a corresponding expense is involved.

Where gas and electricity supplies are concerned, the cost of running a particular service from the main to the property is usually a function of the number of appliances to be used. The more use that you are going to make of the service the less the initial installation should cost. However, it is always advisable to take this matter up with the Regional Gas or Electricity Authority in the initial stages of planning, as a number of factors are involved.

In the case of water, the local Water Board will make a connection to the main and the owner of the plot is responsible for the rest of the installation, subject to the Board's inspection.

One final point which may affect the choosing and purchase of a site. It is advisable to check whether the sale of the land is affected by a Vendor's Clause. This means that the vendors of the land, either a private person or a Local Authority, may have laid down special conditions that must be complied with before there can be any completion and conveyance of the land. For example, it may be that the vendor may have to approve the elevational treatment of the proposed dwelling. You will also, of course, need to obtain planning permission from the Local Authority.

Having decided that the site you have chosen is ideal and that

you want to go ahead and build, a set of comprehensive plans will
be the next requirement.

There are four possible ways of obtaining a plan. Firstly, you
can go to an architect. From him you should obtain a plan *par
excellence,* but you will have to pay for this service. The usual
fee is a percentage of the contract price of the house, and this can
vary depending upon the amount of supervision and other architectural services involved.

The second method is to approach a builder and use one of his
existing designs which he is building at the time. Obviously you
will only use this method when you intend to get the same builder
to put up your house for you, but the cost of the plans will be
much less.

The third method is to have plans drawn up for you by an
architectural draughtsman. Many of them do this as a spare-time
occupation and, of course, no site supervision is included.

The fourth and last method is to draw up your own plans. This
may sound rather frightening at first, but with a little care and
patience it can be done. The author knows of several instances
where this has been done. After all, there is no dearth of published
plans these days, and, although these are copyright, a study of
them will enable you to obtain a good idea of what is required.

The drawing instruments required for a plan of this nature are
quite elementary. A scale which reads $\frac{1}{8}$ in. and $\frac{1}{2}$ in. to the foot, a
set square and pencils and rubber are about all that you need. A
small pair of dividers would be an asset. An essential, however,
is a copy of the Building Regulations 1965. This is published by
H.M. Stationery Office and costs $72\frac{1}{2}$ p. In addition, amendments
are published from time to time and these are available at an
extra cost. A catalogue of window frames (either wooden or steel,
depending on which are to be used) would also be useful, and this
would be obtained from a builders' merchant or direct from the
manufacturers.

If the reader decides to draw up his own plans and specifications
he should read this book through first of all. He will then have a
good idea of the details that go to make up a small house or
bungalow, and how the Building Regulations affect these details.

It is whilst in the process of drawing up the plans for a house that there will be a large number of queries raised, both on the technical aspect of the building itself and on the equipment which is going into that building.

Fortunately, there is no shortage of information which is available for both the technician and the layman. For those who live within reach of London, there is the Building Centre, and this is well worth a visit from anyone who is going to build a new house.

The Building Centre is a permanent but changing exhibition of building materials and equipment, housed in Store Street, off Tottenham Court Road. Manufacturers have facilities for exhibiting a range of their products, and information on them is available at an enquiry bureau. Admission to the Centre is free.

There are also Building Centres in Bristol, Cambridge, Glasgow, Manchester and Southampton which, with London, are all in the Building Centre Group, and there are other independent centres in Birmingham, Coventry, Dublin, Liverpool, Nottingham and Belfast. In addition there is also a Building Information Centre in Stoke-on-Trent.

The main Development Associations are as follows:

Electricity Council (E.D.A. Division), 30 Millbank, London S.W.1.
The Cement and Concrete Association, 52 Grosvenor Gardens, London, S.W.1.
The Copper Development Association, Orchard House, Mutton Lane, Potters Bar, Herts.
Lead Development Association, 34 Berkeley Square, London, W.1.
Timber Development Association, Hughenden Valley, High Wycombe, Bucks.
Zinc Development Association, 34 Berkeley Square, London, W.1.

In addition to the above, a wealth of information is obtainable direct from many manufacturers, and quite often they will even send a representative to discuss matters with you.

By obtaining as much information as possible from various sources during the initial stages, work on the plans will be easier, and the understanding of many details so much greater.

All plans should be drawn on tracing paper so that an unlimited number of prints can be obtained. Usually about a dozen copies will be required.

A plan submitted to the Local Authority has to show the following items, and the usual scale is $\frac{1}{8}$ in. equals 1 ft.

A SECTIONAL PLAN This is a view looking down on the rooms, assuming that the building has been cut through horizontally on a plane whose height is approximately halfway up the windows. For a bungalow only one such plan is needed, but for a house two plans will have to appear, one for each floor. (*Figures 1 and 2*)

ELEVATIONS These are views on the front, back and both sides of the proposed dwelling, and all four should appear (*Figures 3, 4, 5, 6*)

SECTION This is a view that you would get if the building was cut completely through in a vertical plane. This can be either lengthways or breadthways, but it is usual to choose the direction which shows the most roof detail (*Figure 7*)

ROOF PLAN This is a view looking down on the roof timbers with the roof covering removed.

SITE PLAN This is drawn to a much smaller scale than $\frac{1}{8}$ in. to a foot, as it has to show the entire plot, and the minimum scale stipulated in the Building Regulations is not less than 1/1250. However, if planning consent is required in addition to Building Regulation approval (as is nearly always the case) the scale of the site plan has to be 1/500. So, as the scale of 1/1250 is a minimum requirement it is much better to draw the site plan to the larger scale, thus avoiding the need for two plans.

When a scale of 1/500 is used one inch on the plan represents 500 inches on the ground. Thus $\frac{1}{2}$ in. represents 20 ft. 10 in., $\frac{1}{8}$ in. represents 5 ft. $2\frac{1}{2}$ in. and so on.

The house is drawn on this plan as just an outline, and the sewers are indicated and the connections to them. The details of

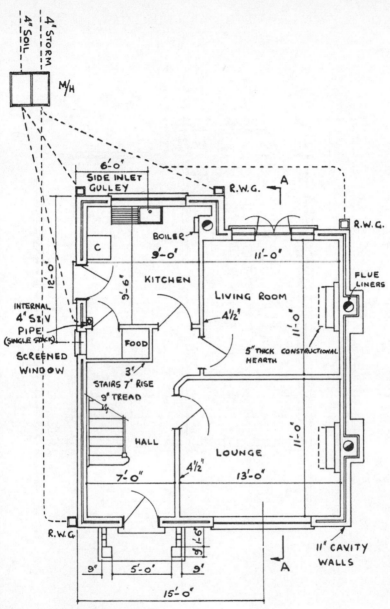

Figure 1 Ground-floor plan

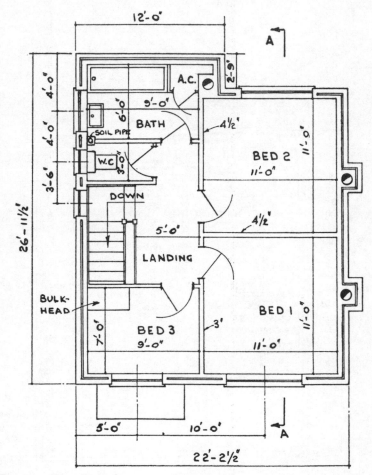

Figure 2 First-floor plan

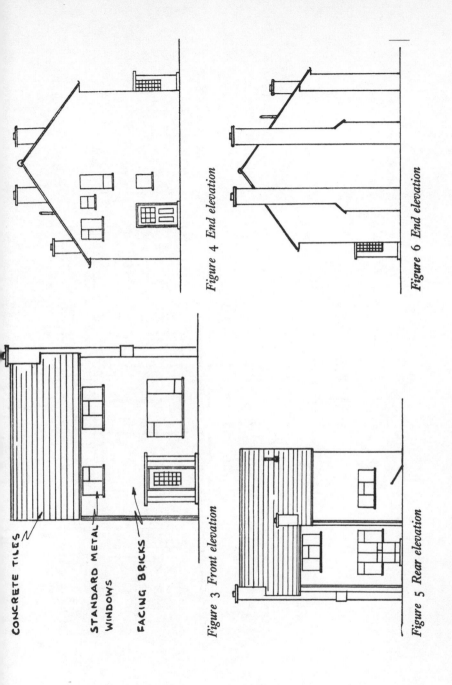

CONCRETE TILES

STANDARD METAL
WINDOWS

FACING BRICKS

Figure 3 Front elevation

Figure 4 End elevation

Figure 5 Rear elevation

Figure 6 End elevation

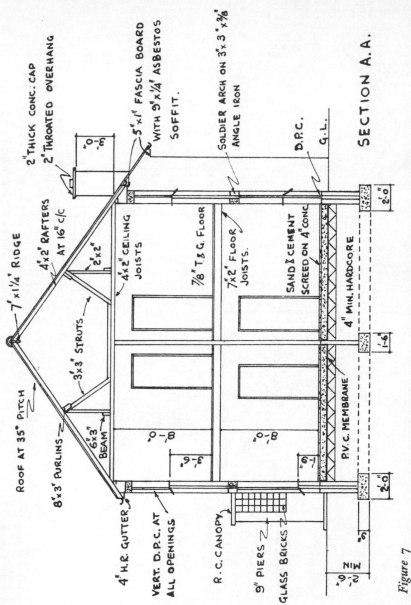

2" THICK CONC: CAP
2" THROATED OVERHANG
5" x 1" FASCIA BOARD
WITH 9" x ¼" ASBESTOS SOFFIT.
4' x 2' RAFTERS AT 16" c/c
7' x 1¼' RIDGE
ROOF AT 35° PITCH
SOLDIER ARCH ON 3" x 3" x 3/8" ANGLE IRON
2" x 2"
4" x 2" CEILING JOISTS
D.P.C.
G.L.
SECTION A.A.
3 x 3" STRUTS
⅞" T.& G. FLOOR
7" x 2" FLOOR JOISTS.
8' x 3' PURLINS
6" x 3" BEAM
SAND & CEMENT SCREED ON 4" CONC.
8'-0"
8'-0"
4" MIN. HARDCORE
4" H.R. GUTTER
VERT. D.P.C. AT ALL OPENINGS
R.C. CANOPY
9" PIERS
GLASS BRICKS
3'-6"
1'-6"
P.V.C. MEMBRANE
2'-0"
1'-6"
2'-0"
9"
9"
MIN
2'-6"

Figure 7

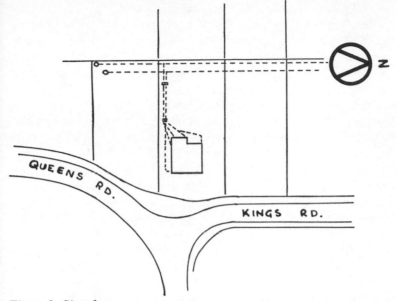

Figure 8 *Site plan*

the site can be obtained from the council offices. If they haven't
a small plan that you can take away it should be possible to trace
the small area required from one of their large plans. This
can then be re-traced on to your own plan and the house added
to it.

When drawing your own plans it is a good idea to rough out the
sectional plan to a larger scale. This could be $\frac{1}{2}$ in. to the foot. By
now you will obviously have plenty of ideas about what you want,
and the study of published plans will show you how various
features look.

The first stage, then, is to rough out a plan to this larger scale.
There will be lots of rubbing out at this stage, but don't worry
about that. This will be the time to decide whether you are going
to have a separate dining room, a dining recess, or a large living-
cum-dining room. Is the hall to be spacious, or are you going to
keep it to a minimum and utilize the extra space on the rooms?

The aspect of the rooms is important. The living room should generally face westwards. The kitchen should look out into the garden, but there are people with a preference for kitchens facing the front. The larder should be on a north wall as this is the coolest. Bedrooms are nice if they can face east, as you then get the morning sun on waking. In three bedroom houses it will obviously not be possible to get this condition with all three, or for that matter get the ideal conditions for some of the other rooms either. A balance will have to be struck. Bathrooms and toilets are best kept to the side or rear so that windows with obscure glass do not intrude on the frontal elevation.

In roughing-out the plan, remember that whilst external walls are 11 in. thick, internal walls will be 3 in., 4 in. or $4\frac{1}{2}$ in., depending whether they are constructed in breeze-blocks or brick. In a bungalow internal walls will be $4\frac{1}{2}$ in. brickwork. In a house internal walls will be $4\frac{1}{2}$ in. brickwork downstairs, and some of these walls will continue on up to the first floor. The other walls upstairs will generally be 3 in. breeze- or hollow-block construction.

Door widths are 2 ft. 6 in. for internal doors and 2 ft. 9 in. for external doors. Smaller doors can, of course, be used for items such as cupboards and larders, but always keep to a standard available size. When contemplating putting doors in short stretches of wall, remember that the 2 ft. 6 in. only embraces the door, and allowances must be made for the frame or lining.

Positions and lengths of windows should be worked out, using a catalogue. With metal windows, standard units can be joined together, both horizontally and vertically, to give a great variety of fixed and opening light combinations.

When planning window positions it is particularly important to ensure that the required zones of open space outside the windows of habitable rooms are preserved. In general terms, these windows are required to overlook land which belongs to the house for a minimum distance of 12 ft.

This regulation effectively positions the house or bungalow within the site and also gives a firebreak from the adjoining land. In connection with this latter point, external walls which are situated within three feet of the boundary should be imperforate, without any windows or doors which are classed (together with certain cladding materials which do not have a specified fire resistance) as unprotected areas.

In point of fact, on a narrow fronted site the windows would, of necessity, be mainly on the front and rear walls, and as the distance from the site boundary increases the area of allowable unprotected area increases also.

Another item which will affect the plan is the staircase. When these details are being worked out it will probably be necessary to draw a local section of the staircase to ensure that the appropriate Building Regulations are covered and that sufficient details are given in the finished plans.

The main requirements are that, between consecutive floors, there must be an equal rise for every step or landing and an equal going for every parallel step. It is also important to note that the sum of the going for a parallel step plus twice the rise must not be less than $22\frac{1}{2}$ in. In no instance may the rise of the step be more than 8 in., the going less than $8\frac{1}{2}$ in. or the pitch of the stairs greater than 42 degrees.

When a wall has to be built over the stairs, care must be taken to see that sufficient headroom is available, if necessary achieving it by a bulkhead. Clearance here should be not less than 5 ft. measured at right angles to the pitch line and not less than 6 ft. 6 in. measured vertically above the same datum. Space under stairs and landings can be utilised for cupboards and larders.

While all these various design points are being given careful consideration, thought should also be given to garage requirements. As mentioned earlier in this chapter, this item can be part of the house structure, built on as the general construction proceeds, or it may be added at a later date.

When the garage is to be built onto the house at a future date the house should be designed with one wall uncluttered by windows or doors. In this way the garage extension may be subsequently added with the minimum of alteration.

Garages adjoining the house must be provided with a half-hour fire protection. A communicating door would be allowed providing that it was self-closing and had a half-hour fire rating. In addition, the floor of the garage should be at least 4 in. below that of the house.

Consequently, even if it is not intended to build the garage at the same time as the house, a little forethought while you are working on the plans could save additional cost in the future. For example, the foundations for a garage and a drain connection could be more easily laid during the early stages of construction than at a later date. If such a connection is included in the early work it must, of course, be laid at the correct depth and properly protected.

When all these points have been settled on the large-scale plan it can then be redrawn at ⅛ in. to a foot on the main plan.

The next view to be worked on should be the section. This gives you such information as cill heights, the height of the eaves and the height of the ridge. This enables the elevations to be completed. As mentioned earlier, the roof members are shown in the section, and at this stage reference will have to be made to the appropriate tables in the Building Regulations to decide sizes of floor joists, ceiling joists, rafters, purlins and ridge. Sizes depend on unsupported spans, and with the exception of the rafters this information is found from the plan. The span of the rafters can be obtained from the section, and will vary with the angle of the roof. The steeper the roof the greater the length of the rafter, and correspondingly the greater the spans between wall-plate and purlin and purlin and ridge. These details will also be needed for the roof plan.

Foundation details are also included in the sectional view and depths of concrete, width and thickness should be indicated. Details of concrete floors and hardcore filling should also be detailed in this view.

When the section has been completed, the four elevations can be tackled. With the information contained on the plan and the section this is relatively a simple matter. The widths and lengths of the house and the window positions are obtained from the plan, and the heights of eaves, ridge and windows are obtained from the section. These elevations are simply a matter of drawing a pictorial representation of the dwelling as seen from four directions.

The addition of the site plan then completes the drawing, and the method of obtaining the information for this has been discussed earlier in this chapter.

The next step will be to obtain a number of prints of your drawing. If there is no printing and drawing office service in your locality, you will have to send away to one of the firms that specialises in this sort of work. It would be as well to obtain about a dozen prints, for, apart from those required by the council, the others will be required by people quoting for various jobs as the construction of your house progresses.

One of the prints submitted to the council is coloured, either by water-colours or coloured pencils.

The following are the correct colours for the different building materials:

Brick	Vermilion
Concrete	Hooker's Green No. 1 Light
Hardcore	Chrome Yellow
Asbestos Cement	Neutral Tint
Tiles	Chrome Orange
Cement Screed	Payne's Grey
Glass	Cobalt (Light Wash)
Wood: unwrot	Raw Sienna
Wood: wrot	Burnt Sienna
Partition Blocks	Discretionary

Two

The Specification

Hand in hand with the plans of your new house goes the specification. This is a detailed account of all the labour and materials required for the erection of a building.

A copy of the specification will be required by the council as a record of the work that has taken place, and also by the body granting your mortgage. The document should be typewritten on foolscap paper, and as several carbon copies will be needed, the exact number should be ascertained before the typing is undertaken.

It will be noticed that the specification is set out in individual trades and is not in the order of building. For example, under the heading 'Concretor' is listed all the work that has to be done, regardless of the fact that the concretor comes in at various stages of the building's progress.

The following is a typical specification for a house, and it must be clearly understood that it is only typical, and that various items will be altered depending on the materials used, site details and whether a house or a bungalow is to be built.

Some of the terms and expressions used may not be fully understood, but by the time this book has been read through, most of them should be apparent.

The specification is given complete and without comment, but each item has been numbered, and notes are given where necessary in subsequent chapters to tie up with these items.

Specification of

Labour and materials for the erection and completion of a detached house, and in accordance with the attached plans, at ......Road,........ (District)...... (Town) for A. Man, Esq.

June 1958

EXCAVATOR

1 *Surface digging and filling*
Remove vegetable soil 9 in. deep and spread and level where directed.

2 *Excavate for drains, etc.*
Excavate to the required widths and depths for all drains, manholes, gas, electric supply and water mains, fill in after laying of same, all filling to be well punned in layers.

3 *Strutting and planking*
The Contractor to strut and plank all excavations necessary to maintain the side of the excavation, to provide all material for so doing, and cart away at completion.

4 *Excavate for foundations*
Excavate to form foundation trenches to conform to the details given on the plans.

5 *Filling-in*
All trenches to be filled in as soon as walls are above ground level. All filling to be well punned in layers. Spread and level surplus on site where directed.

6 *Hardcore*
All ground-floor solid floors to be laid on 4 in. minimum of approved dry broken brick or concrete, broken to a 3½ in. gauge, levelled, well rammed, and blinded with fine materials to receive

B

concrete. No excavated materials from site are to be used for this filling.

CONCRETOR

7 *Cement*
The cement used is to be Portland cement of approved British manufacture of a quality which complies with the requirements of specification No. 12 of the British Standards Institution.

8 *Sand*
The sand is to be good clean, sharp, washed river or pit sand, free from organic or earthy matter.

9 *Aggregates*
The aggregates for various mixes of concrete are to be clean screened gravel of the size stated, and to comply with B.S. 882.

10 *Concrete*
The various mixes are to be as follows:
For foundations: 1 of Cement, 8 of 'all in' aggregate, $1\frac{1}{2}$ in. to $\frac{3}{16}$ in., well graded. (Mix A.)
For solid floors: 1 of cement, 6 of 'all in' aggregate, $1\frac{1}{2}$ in. to $\frac{3}{16}$ in., well graded. (Mix B.)
For reinforced work: 1 of cement, 2 of sand and 4 of aggregate $\frac{3}{4}$ in. to $\frac{3}{16}$ in., well graded. (Mix C.)

11 *Mixing and placing*
The materials for all concrete are to be gauged by actual measurement and are to be mixed on a clean close-boarded platform, turned over three times in a dry state so that the cement is evenly distributed, and afterwards mixed with a sufficient quantity of clean water applied through a rose to produce the required workability, turned over twice and then to be immediately deposited in trenches, or well spread and filled in around filler joints or reinforcing rods and beaten down and levelled on the top.

Alternatively the concrete may be mixed by an approved

mechanical mixer in small batches, to a uniform consistency for a period of not less than one minute.

Only such quantities as are required for immediate use are to be mixed at any time, and then carefully deposited within the limit of the initial set for concrete (i.e. 45 min.).

12 *Jointing*
Where fresh concrete is to be deposited against concrete already set or hardening, the existing surface is to be hacked, cleaned, wetted and covered with cement mortar (1 : 2) not less than ½ in. thick, and the fresh concrete well rammed against the surface.

13 *Frost*
No concrete shall be mixed or laid when the temperature is 38° F., or below.

14 *Protection*
All concrete must be protected by straw and sacking as necessary, and no frozen aggregate or sand shall be used. A curing period of seven days should be allowed.

15 *Foundations*
Foundations to walls to be formed of concrete as specified (mix A) to the widths, and thicknesses shown in the drawing. Remove levelling pegs as concrete is laid.

16 *Damp-proof membrane under floors*
Provide and lay continuous membrane of approved PVC sheeting under all solid concrete floors.

17 *Solid floors*
Lay on hardcore previously specified 4 in. bed of concrete as specified (mix B) spread and levelled and finished with a tamped finish to receive a cement-and-sand screed.

18 *Concrete lintels*
Provide and set over door and window openings precast concrete lintels. Lintels are to be 9 in. deep and have a 9 in. bearing at each

end. Reinforcement to consist of one $\frac{1}{2}$ in. mild steel rod for openings up to and including 4 ft. and $\frac{3}{4}$ in. mild steel rods (one for each $4\frac{1}{2}$ in. and part thereof of the thickness of the lintel) for openings over 4 ft. Rods are to be hooked at both ends and embedded 1 in. from the bottom. Cast in lintels dovetailed hardwood plugs where required for joinery fixings—these are to be at least $\frac{1}{4}$ in. from reinforcement. Faces of lintels to be rendered or plastered are to be hacked and roughened for key. 4 in. × 3 in. × $\frac{3}{4}$ in. steel angles are to be fixed to support outer skins of cavity walls over all openings.

19 *Concrete canopy*
Provide reinforced concrete canopy and lintel over front door, cast as plans, and reinforced with $\frac{3}{4}$ in. diameter mild steel rods. Rods to be hooked at both ends.

20 *Larder slab*
Form 2 in. precast slab reinforced with No. 8 B.R.C. Fabric finished smooth on all faces.

21 *Chimney cap*
Provide formwork and cast $2\frac{1}{2}$ in. thick weathered fine concrete caps to chimneys. Caps to be waterproofed and perforated to receive flue lining. Form throating round soffit of slab.

22 *Concrete steps*
Form steps in concrete (mix B) to entrances where shown on drawings, finished for, and floated with granolithic paving, within three hours of placing the base concrete.

23 *Mortices*
Form all mortices and holes required for pipes, etc., for all trades as required in concrete and reinforced concrete.

24 *Shuttering*
Shuttering, where required, to be of substantial character and well strutted and braced.

DRAINLAYER

25 *Sand, cement, concrete, etc.*
The cement, sand, ballast, bricks and other materials are to be as
described in the various trades.

26 *Stoneware pipes*
The stoneware pipes and fittings are to be best quality salt-glazed
stoneware, to comply in every respect with British Standard
Specification No. 65 and the B.S.S. mark shall be indicated on
these pipes; the pipes are to have socketed joints caulked with
tarred gaskin and to be joined in cement and sand 1 : 1.
The inside of the pipes are to be carefully cleaned out and
cored with a disc on rods so as to leave clear an unobstructed
waterway. The drain fittings, channels, junctions, bends, traps
gullies, etc., to be salt-glazed ware to B.S.S. No. 539 if applicable.

27 *Laying*
Pipes are to be laid as shown on the plans in straight runs and to
even gradients, or not less than 1 in 40 for 4 in. pipes and 1 in 60
for 6 in. pipes, all pipes being laid on even solid foundations for
the full length of each pipe, and haunched in weak concrete 1 : 12.
Where required by the nature of the soil, pipes to be laid on
concrete 1 : 2 : 4 (mix A), in all cases to be in accordance with the
Building Regulations and to the satisfaction of the Local
Authority.

28 *Testing*
After the drains have been laid and the necessary work of haunching, surrounding with concrete and backfilling completed, they
are to be subjected to a water test in the presence of a representative of the Local Authority. The drains must be thoroughly
sound and watertight and any sections not meeting these requirements to be taken up and relaid at the contractor's expense.

29 *Fillings*
In filling the trenches after the drains have been laid great care
is to be taken so as not to disturb the drains, and the best of the

excavated materials is to be used for packing round the pipes.
The whole is to be carefully and thoroughly consolidated, and the
ground levelled up for hardcore and pavings.

30 *Manholes*
Build manholes at positions shown on the drawings to sizes and
depths required with 6 in. thick, 1:2:4 concrete base (mix B)
projecting 4 in. all round, and with 9 in. thick walls of common
bricks in cement mortar. Lay 4 in. diameter salt-glazed stoneware
channels and connect to drain, and lay ¾ in. channel branches all
set and jointed in cement and neatly flaunched round. Form
fine concrete benching average 3 in. thick worked around channels
and branches and render sides of manholes and benching in
cement and sand 1:1 and trowel smooth.

31 *Manhole covers*
Manhole covers and frames to be 24 in. × 18 in. cast iron com-
plying with B.S.S. No. 497, and minimum weight of 56 lb.
   Bed frames in cement and seal covers in cart grease and sand.
Render top of manhole walls and level off to cover and ground.

32 *Gullies*
Provide and set, where shown on drawings, approved pattern
glazed stoneware trapped gullies and covers bedded on and surrounded in concrete.

33 *Large radius bend*
Provide and fix 4 in. diameter large radius salt-glazed bend at the
foot of the soil pipe.

34 *Connection with sewer*
Give notice to the Local Authority and pay their fees for making
connection with the sewer.

BRICKLAYER

35 *Bricks*
The whole of the bricks are to be good, hard, sound, whole and

well-burnt bricks, free from flaws and to give a clear ring when
struck against another brick. To be 2⅝ in. nominal depth.

36 *Flettons*
The bricks for general building purposes are to be commons
obtained from the local brick company.

37 *Facing bricks*
Facing bricks to be best quality local brick company facings and
to be square and true, or other approved.

38 *Sand and cement*
The cement is to be as specified under 'Concretor,' and the sand
to B.S. 1200, or other approved.

39 *Cement mortar*
The cement mortar is to be composed of one part by measure of
Portland cement and four parts of bricklaying sand. The cement
mortar is to be mixed in small batches, as required for use, and
used immediately after mixing. A proper stage is to be provided
for mixing and the water must be clean and added from a can with
a fine rose.

40 *Dimensions*
All brickwork, walling and partitions are to be set out and built
to the respective dimensions, thicknesses and heights shown on
drawings.

41 *Wetting bricks*
All bricks are to be well wetted before being laid and the tops of
walls where left off are to be well wetted before resuming work.

42 *Frost*
No brickwork is to be carried out when the temperature is below
38° F. and all new laid brickwork is to be properly protected from
frost and covered up with boards or sacking.

43 *Hollow walls*
Hollow walls shown on the plan to be built in two half-brick thicknesses with a 2 in. cavity bonded together with mild steel, zinc-coated, knife-edged-type, wall-ties to B.S.S. No. 1243. The cavity is to be kept clear of all rubbish and mortar droppings by movable boards or other means. The wall-ties are to be carefully laid and in no cases are they to fall towards the inner thickness of the wall, and all mortar droppings are to be cleaned off as the work proceeds.

44 *Flush up*
All joints are to be thoroughly flushed-up as the work proceeds.

45 *Openings, traps and D.P.C. window cills*
Build in above all door and window openings to cavity walls approved Bituminous D.P.C. 2 in. beyond each lintel, turned 1½ in. into cavity on inner leaf, and dressed-down ¼ in. over surface of door- or window-frames, and also all cills, and build 4½ in. into brickwork at end.

46 *Facing*
Face all external walls and stacks with facing bricks. Bricks previously specified.

47 *Jointing*
All external brickwork to be jointed with a neat recessed joint as the work proceeds.

48 *Damp-proof course*
Build in all walls and chimney breast, and to full thicknesses thereof, bituminous sheet D.P.C. complying to B.S. 743.

49 *Window cills*
External cills to be Blue Broseley Tiles set at 45 degrees, projecting 2 in. from face of brickwork extending 6 in. both sides of window opening. Internal cills to be in hardwood except as specified below.

Bathroom cill in black Vitrolite.
Kitchen cill in white glazed tiles.

50 *Fixing steel casements*
Fix, including building in lugs, standard steel casements by an
approved maker, to sizes and types shown in the drawings.

51 *Brickwork in cement mortar*
All brickwork in chimney-stack above roof level to be built in
cement mortar.

52 *Chimney opening*
Build in over fireplace opening reinforced concrete lintel. Gather
in and form throat to flue immediately above opening.

53 *Flues*
Flues to be constructed and lined with approved rebated or
socketed clay flue linings to B.S. 1181.

54 *Jointing*
Flue linings are to be built in with socket uppermost and are to
be jointed and pointed with cement mortar as the work proceeds.

55 *Chimney caps*
Finish tops of chimney-stacks as detailed on plans, and terminate
flue with 9 in. diameter terra-cotta, dwarf chimney-pots, set and
flaunched in cement.

56 *Rough render*
Rough render all breasts and stacks where they pass through roof
and floor.

57 *Ventilating gratings*
Provide and build in 9 in. × 3 in. air bricks for ventilating cavity
as detailed on drawing.

58 *Build in*
Build in all door-frames in mortar, and bond in metal ties with
brickwork. Point to match facing.

59 *Fixing bricks*
Provide and build in coke breeze fixing bricks, or hardwood plugs, as required for fixing joiners, and other work.

60 *Rake wedge and point*
Rake out joints of brickwork for all lead stepped and cover flashings, aprons, etc., wedge with led wedges and point to match facings.

61 *Clean down*
Point up around frames, etc., and clean down all external brickwork on completion.

62 *Fireplace*
Allow the P.C. sum of £35 per tile or brick fireplace, and fix, including all necessary cement, fireclay, fine concrete filling behind fireback and laying hearth, to be approved by building owner.

CARPENTER AND JOINER

63 *Defects*
Any portions that warp or develop shakes or other defects to be replaced before being wedged.

64 *Roof*
Construct the roof as shown on the drawing with timber of the sizes specified below, framed together in a proper workmanlike manner, the rafters spaced not more than 16 in. centre to centre, notched and spiked to ridge boards, purlins and ceiling joists.
Rafters 4 in. × 2 in.   Purlins 8 in. × 3 in.
Ridge   7 in. × 1 in.   Struts 3 in. × 3 in.
Plates 4 in. × 2 in.
Binders 5 in. × 3 in.   Ceiling Joists 3 in. × 2 in.

65 *Felt to roof*
Cover the roof with a layer of approved roofing felt not less than $\frac{1}{16}$ in. thick, nailed to rafters and lapping 4 in. at horizontal joints.

66 *Eaves*
Fix 5 in. × 1 in. wrot deal fascia to eaves fixed to rafter feet and
provide and fix ¼ in. asbestos soffit screwed to bearers.

67 *Ceiling joists*
Provide and fix 3 in. × 2 in. ceiling joists at not more than 16 in.
centres.

68 *Trimming*
Properly trim as required for trapdoor and chimney-stack.
Trimmed and trimmer to be of same scantlings, unless otherwise
stated.

69 *Fixing of joiner's work*
All joiner's work generally to be secured to plugs built into
brickwork.

70 *External doors*
External doors to be B.S.S. and obtained from an approved
supplier and according to the accompanying schedule.

71 *Internal doors*
Internal doors to be flush doors, plywood faced, obtained from an
approved firm, and according to the accompanying schedule.

72 *Cupboards*
Provide the necessary framing for linen cupboard, fit, secure and
effect necessary stopping.

73 *Skirtings and architraves*
Provide and fix ¾ in. × 4 in. rounded deal skirtings on walls,
and ¾ in. × 2 in. round deal architraves to all internal door
openings, all securely plugged to walls.

74 *Trapdoor*
Trim ceiling and provide and fix 1 in. wrot deal trapdoor, size
2 ft. 6 in. × 2 ft. 6 in. clear, with wrot deal, rebated and beaded
linings.

75 *Shelves in linen cupboard*
Provide and fix three tiers of shelves, consisting of wrot deal slats
2 in. × ¾ in., spaced 1 in. apart on cross battens, screwed with
brass screws and resting on 2 in. × 1 in. bearings securely plugged
and fixed on the walls of the cupboard.

76 *Shelves in larder*
Provide and fix in larder three tiers of 9 in. × ¾ in. shelving as
directed.

77 *Door furniture*
Internal doors are to have mortice locks and approved furniture.
Main entrance door is to have a Yale lock, two barrel bolts and
approved furniture.

78 *Bath panel*
Frame to one side of bath with 2 in. × 1½ in. framing, with
uprights at 15 in. centres, and face with hardboard of approved
manufacture.

79 *First-floor joists*
The joists are to be of the dimensions and spacings as shown on
the drawings, and shall be laid in parallel lines. Where they extend
into the thickness of walls, they shall be brush creosoted on all
surfaces.

80 *Herring-bone strutting*
All joists over 8 ft. span to have one row of 2 in. × 1½ in. herring-
bone strutting.

81 *Stud partitions*
Provide 3 in. × 2 in. stud partitioning where shown on drawing,
the whole to be properly framed and securely fixed between floor
and ceiling. Frame members adjacent to walls shall be securely
anchored to them.

82 *Floor-boarding*
Floor-boards shall be tongued-and-grooved and to the full depth shown on drawings. They shall be securely fixed with 2 in. nails at each fixing position on joists. They shall be laid in as long lengths as possible and heading joints shall be evenly distributed and framed over joists. Skirtings shall not be installed until after the floor-boards are fixed.

83 *Staircase*
Form staircase in position shown on plan. All overall dimensions are to be taken on site after walls and floors have been finished. The following sizes shall be strictly adhered to: Newel posts 3 in. × 3 in., wall stringers $1\frac{1}{4}$ in. thick, centre stringers $1\frac{1}{2}$ in. thick, treads 1 in. thick, risers $\frac{3}{4}$ in. thick, balusters 1 in. × 1 in. and handrail 2 in. × $1\frac{1}{2}$ in. Form landing with tongued-and-grooved boarding on 4 in. × 2 in. bearers tenoned to 7 in. × 3 in. trimmer.

PLUMBER

84 *External sheet lead*
All sheet lead to be the best milled lead free from all defects, and of even thickness, and of the several weights specified and laid and dressed in the best manner.

85 *Laps, tacks, etc.*
All laps in flashings, aprons, drips, etc., are to be 4 in. wide and in stepped flashings 6 in. wide. All leadwork to be secure with 2 in. lead tacks, and all nails used to be copper.

86 *Chimney aprons*
Provide and fix to upper sides of chimney-stack 4 lb. lead aprons, 12 in. girth, turned up the vertical face of the brickwork 6 in. and let $\frac{3}{4}$ in. into joints and securely fixed with lead wedges.

87 *Cover flashings*
The lead aprons to stacks to have 5 lb. lead cover flashing secured as above.

88 *Internal pipes*

Internal pipes are to be of copper tubing of the following gauges, to B.S.S. 659, with compression joints:

$\frac{1}{2}$ in. diameter pipes	18
$\frac{3}{4}$ in. diameter pipes	18
1 in. diameter pipes	18
$1\frac{1}{4}$ in. diameter pipes	17
$1\frac{1}{2}$ in. diameter pipes	17

89 *Fixing of pipes*

All pipes are to be run on the surface of walls and fixed with approved pipe clips.

90 *Connection with main*

Arrange with the Local Water Authority for connection to their main, and pay their charges for doing so.

91 *Stop-cocks*

Provide a $\frac{1}{2}$ in. stop-cock to the main supply pipe in a convenient position with an approved cast-iron casing with a hinged locking cover, and set in a stop-cock pit and make good surfaces of paving around. Provide and fix $\frac{1}{2}$ in. approved high pressure gun-metal stop-cock at lowest part of rising main.

92 *Main supply pipes*

The contractor is to provide for a $\frac{1}{2}$ in. soft copper supply pipe, laid in a trench not less than 2 ft. 6 in. deep, and allow for filling in and making good.

93 *Rising main*

From the main supply pipe carry $\frac{1}{2}$ in. rising main to bathroom.

94 *Drinking points*

From rising main take $\frac{1}{2}$ in. branch to sink in kitchen and fix high-pressure, easy-clean, chromium-plated bibcock over sink.

95 *Bath*

Supply and fix, where shown, approved 72 in. porcelain cast-iron bath with front panel, including hot and cold pillar taps and waste with 1½ in. union and deep seal antisyphonage trap, and 1½ in. overflow. Provide and fix 1½ in. waste-pipe and carry through wall to discharge into gully as shown on plan.

96 W.C.
Supply and fix in lavatory, a white fireclay pedestal closet set with a plastic seat, with low-level, waste-water preventer, to be approved.

97 Kitchen sink
Provide and fix sink unit with stainless steel bowl and drainer, and complete with waste fitting, plug, chain and 1½ in. union. The contractor to allow for providing and fixing 3 in. deep seal anti-syphonage trap with screw cleaning eye. The waste pipe to be taken through the wall to discharge into side inlet gulley as shown on plan.

98 Cutting away and making good
Cut and pin all ends of pipe clips and brackets, cut holes in walls, floors and ceilings for all plumber's work, and make good around all trades.

99 Testing
All tanks and fittings to be tested by the Local Authority at the Contractor's expense. Test the whole of the internal plumbing work, water supply and fittings, and leave in perfect working order at completion.

100 Soil, waste and vent pipes
Soil, waste and vent pipes to be of approved PVC materials of adequate strength and durability. Vent pipe to have suitable weather-proof flashing where it goes through the roof and to be topped with a galvanised wire balloon cage.

101 Joints
All joints to be made to manufacturer's instructions and to the approval of the Local Authority.

102 *Fixing*

All PVC soil and vent pipes to be securely attached to the internal wall with 4 in. ring clips spaced and fixed to manufacturer's instructions.

103 *Soil branch*

From W.C. take 4 in. W.C. connection and fit to curved branch on soil pipe.

104 *Rainwater pipes*

Rainwater pipes and fittings to be PVC $2\frac{1}{2}$ in. diameter, circular pattern, jointed with Neoprene gaskets to manufacturer's instructions. They are to be attached to external walls with approved pipe clips with galvanised mild steel backplates.

Provide all outlets, swan necks, bends and other fittings required.

105 *Eaves gutters*

Eaves gutters to be PVC 4 in. diameter half-round eaves gutters, joined with gutter straps and Neoprene gaskets to manufacturer's instructions. Moulded PVC fascia brackets to be fitted every 6 ft.

106 *Hot-water system*

Instal boiler of approved manufacture in kitchen, with necessary connections and pipework to hot-water tank in airing cupboard and storage tank in roof.

TILER

107 *Tiles*

Cover roof with $16\frac{1}{4}$ in. × 13 in. variable gauge interlocking tiles, surfaced with permanently coloured granules of light red colour.

108 *Felting*

Cover rafters with suitable quality felt, lapped 3 in. at horizontal joints and 6 in. at vertical joints. The felt to sag slightly between rafters and carried over fascia board to give drip into gutter.

1. Profile boards

2. Foundation trench, showing pegs and steps

3. Foundation trench, after concreting

4. Solid floors in process of being laid. Note tamping-board

5. Door-frame being built in. Note vertical D.P.C. where cavity is being closed

6. Cavity being closed at window-frame. Note vertical D.P.C.

7. Internal walls under construction, showing fireplace opening and lintel

8. Soldier arch and supporting angle. The wall-ties are for an attached lintel, but a separate lintel on the inner skin would be better practice

109 *Battening*
The battens to be of good quality reasonably free from knots and laid to a gauge specified by the Tile Company.

110 *Tiling*
Lay tiles specified to lap as instructed by Tile Company. Tiles to be secured every second course with 2 in. galvanised nails.

111 *Eaves*
The first course of tiling to be fixed to project 2 in. over the fascia. Each roll of tile to be bedded on the felt above fascia board.

112 *Verges*
Lay undercloak of plain tiles at verges projecting at least 2 in. over wall. Bed and point tiles on same.

113 *Ridges*
Cover ridge with hog-back ridge tiles to match main roof. Tilt end ridge tiles and fill exposed ends with cement mortar and set in pieces of plain tile.

114 *Mortar*
All bedding and pointing mortar to consist of three parts clean sharp sand to one part of Portland cement.

ELECTRICIAN

115 *General*
Electrical installation to include running all services, fixing meters, main switches, etc., all to be in accordance with the I.E.E. regulations, and the requirements of the Local Electricity Board, and according to schedule, and as shown on drawings.

116 *Main switchgear*
To consist of A.C. 30 amp. to control power plugs on ring main system, and switch splitter, lighting to be split on two circuits.

117 *Wiring*
All wiring to be carried out in tough rubber sheathed, or P.V.C. sheathed cable, either twin with earth, or three core as required. Where cable is buried in plaster, it is to be protected with standard steel channelling.

118 *Power sockets*
Power Sockets to be B.S.S. 13 amp. capacity.

119 *Lighting switches*
To be flush type of approved quality.

120 *Attendances*
The Contractor to allow for attendances, cutting away and making good for all trades.

PLASTERER

121 *Internal plastering*
The whole of the work to be executed in Hemi-Hydrate Gypsum Plasters, and used in accordance with the manufacturers' printed instructions.

122 *Sand*
The sand to be clean, sharp river or pit sand free from loam and other impurities, and should conform to the revised British Standard Specification for Plastering Sand (B.S. 1198); Type One.

123 *Measuring and mixing*
All plaster and sand must be measured by volume, and mixed thoroughly on a clean platform (or mechanical mixer) in a dry state before water is added. The water is to be clean and drawn from a tap or hydrant.

124 *Ceilings and stud partitions*
Fix $\frac{3}{8}$ in. thick Gypsum based plasterboard to all ceilings and stud partitions, with bound edges across the joists. They shall be nailed

with $1\frac{1}{4}$ in. × 12 S.W.G. galvanised or sheradised nails driven in at right angles to the face, care being taken not to break the paper cover. Nails shall not be closer than $\frac{1}{2}$ in. to the bound edges and cut ends. The nails shall normally be 6 in. apart. Continuous joints over the supports shall be avoided by staggering the boards.

125 *Joint scrimming to plasterboard*
All joints and angles to be scrimmed, using neat Gypsum Board Finish Plaster, and $3\frac{1}{2}$ in. jute scrim.

126 *Plastering walls*
All brick walls to be floated $\frac{1}{2}$ in. thick with 1 part Gypsum Browning Plaster to 3 parts sand by volume, laid on evenly, straightened with the rule, and lightly scratched to receive the finishing coat. On breeze or clinker block walls the sand is to be reduced to 2 parts. The finishing coat to be Gypsum Wall Finish plaster, applied neat. The surface to be trowelled to a smooth but not polished surface.

127 *Plastering ceilings and stud partitions*
All plasterboards to be finished with one coat of neat Gypsum Board Finish Plaster to a thickness of not less than $\frac{3}{16}$ in.

128 *Sundry labours*
Perforate plasterboards for pipes and make good as required.

129 *Floor screed*
Lay a 1–4 cement and sand screed, 1 in. thickness minimum for thermo-plastic tiles. The screed to be laid level and true, and finished to a high-class, steel-trowel finish.

GLAZIER

130 *General*
All glass to be best of its kind, free from all defects, and uniform in thickness, well puttied and sprigged where necessary. All putty is to be carefully trimmed and cleaned off to sight lines.

131 *Clear sheet glass*
Glaze the whole of the window, except where otherwise specified, with 24 oz. sheet glass, set in gold size putty.

132 *Obscured glass*
Glaze the windows of W.C. and bathroom with arctic or other obscured glass. Glaze the front and rear entrance doors as shown on plan with similar obscured glass.

PAINTER

133 *Paint*
The whole of the paint is to be obtained from an approved manufacturer, ready mixed for use and delivered in sealed cans, and is to be used in accordance with the maker's instructions.

134 *Washable distemper*
The washable distemper is to be obtained from an approved manufacturer, and is to be mixed in accordance with the maker's instructions, and finished to approved tints.

135 *Specimen*
The Contractor shall provide specmen of any colours or tints or paint, stains and distemper for approval.

136 *Priming joinery*
All joinery required to be painted is to be knotted and primed at the shop before delivery to the site.

137 *Whitening internally*
Prepare and twice-whiten all ceilings to an approved tint.

138 *Distempering internally*
Prepare and twice-distemper to approved tints, all plastered and rendered walls in lounge, hall, and dining room, passages and bedrooms.

139 *Woodwork internally*
The whole of the interior woodwork including fittings is to be
carefully stopped and painted two coats of oil colour to approved
tints.

140 *Pipes internally*
All pipes where exposed are to be painted two coats of oil colour
to match walls and ceilings.

141 *Rubbing-down woodwork*
The surfaces of all work required to be painted are to be well
rubbed down before painting is commenced, and before the
finishing coat is applied.

142 *Woodwork externally*
The whole of the external wrot woodwork is to be carefully
stopped with hard stopping and painted two coats of oil paint to
an approved tint.

Schedule of lighting and power points

	LIGHTING		POWER
	No. of points	*No. of switches*	13 *amp sockets on ring main or spur*
Hall	1	2 (1 two-way)	1
Lounge	3	2	2
Kitchen	1	2	1
Dining room	1	1	2
Landing	1	1 (two-way)	1
Bedroom 1	1	1	1
Bedroom 2	1	1	1
Bedroom 3	1	1	1
W.C.	1	1	—
Bathroom	1	1	—
External	1 over kitchen door		—

Schedule of doors

Door	Size	Frame and linings	Hinges	Furniture
Main entrance	6' 6" × 2' 8"	4½" × 3" rebated and rounded	4" cast-iron butts	One 6" bolt and shoot, letter plate and knocker, mortice lock, 2 keys as approved
Side entrance	6' 6" × 2' 8"	,,	,,	One 6" bolt and shoot, mortice lock and furniture as approved
Hall-Kitchen	6' 6" × 2' 6"	1" lining with 1½" × ½" stop planted on	3" butts	Approved furniture
Hall-Lounge	,,	,,	4" rising butts	,,
Hall-Dining room	,,	,,	3" butts	,,
Landing-W.C.	,,	,,	3" butts	,,
Landing-Bathroom	,,	,,	,,	,,
Landing-Bedroom 1	,,	,,	,,	,,
Landing-Bedroom 2	,,	,,	,,	,,
Landing-Bedroom 3	,,	,,	,,	,,
Kitchen-Pantry	6' 0" × 2' 0"	1" lining with 1½" × ½" stop planted on	,,	Ball catch
Linen cupboard	6' 6" × 2' 0"	2" × 2" rebated	,,	Furniture as above

Three

Foundations—marking out—digging and concreting

Building Regulation requirements

As stated in Chapter 1, a copy of the Building Regulations is a necessity, and a copy may be obtained from H.M. Stationery Office. These Regulations, which came into operation on 1st February, 1966, have replaced the local Building Bye-laws, thus making a standard code which is statutory throughout England and Wales. An exception is made to parts of the London area which come under the London Building Acts, and also to Scotland, which is covered by The Building Standards (Scotland) Regulations 1963.

Before dealing with the Building Regulations which affect this chapter, mention should be made of a particular regulation which affects this and several other chapters. This is the regulation involving the commencement and completion of certain stages of work, and the various sections deal with the written notice required by the Local Authority when the building under construction has reached a certain point. These include the commencement of building, foundation excavations, damp-proof courses, oversite concrete, laying of drains and backfilling of drains.

Usually this is taken care of automatically, as most authorities issue printed cards or forms at the time that the plans are passed, and one of these is sent in at each of the stages mentioned.

When the appropriate notice is given the Building Inspector

surveys the work and permission is given to proceed. It should
also be realised that these are the statutory notifications, and that
the Building Inspector may visit the site at any reasonable time in
order to satisfy himself that the work being undertaken complies
with all the Building Regulations.

Other sections of this regulation deal with giving notice of
completion and prior to occupation, and the taking of samples of
materials by the Local Authority during the construction of the
building.

As far as the foundations are concerned, the requirements are
that they shall be designed and constructed to take the load that
the building superstructure imposes upon it, and taken down far
enough to guard against climatic alterations to the subsoil, clay
being one of the worst offenders as far as the latter item is concerned.

Data is given for the widths and depths of foundations in various
types of subsoil. Generally speaking, the private house of two
storeys exerts a load of 1 ton per lineal foot on the foundations, and
a bungalow three-quarters of a ton per lineal foot. The widths of
the foundations are therefore adequate if taken as being equal to
the width of the wall plus 12 in. The thickness is usually equal to
9 in., but the thickness of foundations for the internal, non-load
bearing walls of a bungalow may be 6 in.

The mix is also specified and can be a minimum of 1 to 9,
one part of cement to 9 parts of graded aggregate. A more usual
mix is, however, 1 to 8. Should the site be sloping, and the
foundations at more than one level, the regulations call for a
minimum distance by which the upper level should extend over
the lower level.

Information and detail on plans

House plans do not usually show the foundations in any views
except the section. In this view the portion of the house which
extends below ground level is shown, and this will mean showing
the concrete foundations as they appear under the walls. Depth of
the concrete below ground, width, and thickness should be

dimensioned on the drawing, and any doubts or queries on these items should be discussed with the Building Inspector.

Constructional details

With the passing of the plans, the actual construction of the house begins, and the first job is setting out. This may or may not be done by the reader, but if you have successfully completed your plans there is no reason why you should not do this job as well.

The purpose of marking-out is to give lines from which the foundations can be dug, and all the walls erected. These lines are obtained by profile boards. These are horizontal boards with pegs at each end. The pegs are knocked into the ground and saw cuts are made on the horizontal portion for trench and wall extremities. When strings are stretched between profile boards at either end of a proposed trench or wall, the boundaries are clearly indicated over the intervening space.

The first step in marking out a small building is to set out the building line. This is done by stretching a length of string (from suitable pegs) across the plot. In special instances this line may vary if you have a corner site or have to line up with existing buildings. It is p rmissible to build behind the building line, but not in front of it.

Assuming that the house is to be on the building line, and this is strung across the plot, the rest of the marking-out will be constructed from this line. It is best to mark out a complete outline of the building in pegs and string. It is then a fairly easy matter to see where the main profile boards will be required, and also those for the internal walls.

The distance between one front corner of the house or bungalow and the boundary should be measured along the building line, and a peg knocked in to represent this point. When one peg has been knocked in, the overall length of the building can be measured and a further peg placed in position.

These pegs should be stout ones, about 2 in. square and sharpened at one end. A nail can be knocked into the flat end and this forms a perfect station to work from.

With the two front pegs in position, each of the back pegs can be positioned by measuring and marking both the depth of building and the length of the diagonal across to the opposite front peg. A check dimension can then be made from the back pegs to either boundary.

When the main pegs are in position, the outline can be indicated with string. As a further check the sides can be tested for being at right angles to one another by the 3, 4, 5 triangle method. This is done by measuring from a corner 3 ft. along one side and 4 ft. along the side at 90 degrees to it. The line completing the triangle (the hypotenuse) should be 5 ft. long if the sides are truly at right angles. Any convenient length of side can be used for the check, providing the ratio of 3, 4, 5 is maintained.

When the outline is complete the profile boards are knocked into place. These must be long enough to embrace the full width of the foundation trench. Sometimes, as in the case of a building that has an L shape, it will be found that by making a corner profile board longer, it can be utilised for an internal wall as well.

Profiles should be knocked into position well clear of the corners, otherwise they will hamper the digging-out.

The next operation is to transfer the outline of the building to the profile boards. A length of string is laid over two corner pegs and extended out to the profiles. When the string is held taut and is directly over the datum nails in the pegs, the position where it strikes the profiles can be marked. These marks should be made permanent by small saw-cuts on the profiles.

The positions of the cavity, the inner skin and the foundation trench can be obtained by direct measurement along the profile, and saw-cuts made in the appropriate place. A position $4\frac{1}{2}$ in. in from the first mark will give the outer skin, a further 2 in. will give the cavity and a final $4\frac{1}{2}$ in. will indicate the inner skin. Six and a half inches either side of the marks representing the inside and outside of the 11 in. wall will give the trench positions.

When all the outside walls have been recorded on the profiles in the foregoing manner, the profiles are put down for the internal walls. Pegs are used to indicate their position first of all, and obtained by direct measurement from the corner pegs already

established. The marking on the profiles is then done in exactly
the same way except that it is not quite so complex, as the wall
thickness is only a single skin of 4½ in. brickwork.

When the marking-out has been completed, a card (already
provided) is sent to the Building Inspector's department, and the
work must be passed by them before any further work is done..

The digging of the foundations can commence as soon as the
marking-out has been passed by the Building Inspector. To mark
one trench for digging, two lengths of string are stretched between
the cuts representing the edges of the foundation trench on two
opposite profile boards.

Vertical spade-cuts are then made along each of the strings for
the entire length of the trench. This enables the turf to be more
easily removed, and this, together with the top soil, should be carefully removed and barrowed away for future use. The subsoil from
these foundation trenches should also be barrowed away. This may
be used for any subsequent building up or terracing of the garden,
or it may be collected by lorry and disposed of. If the latter is to
be the case, the subsoil should be dumped where there will be
easy access for carting away.

When all the external trenches have been dug, short lengths of
planks will be required to cross them, to enable the soil from the
internal trenches to be barrowed away.

When digging trenches, care should be taken to see that the sides
are kept vertical, and, when the finished depth is arrived at, the
bottoms smooth.

If the site is sloping, the digging should commence in the lowest
corner, and in order to prevent the depth of the foundations being
out of all proportion, they will have to be stepped. In this way, the
final depth of the trenches is not very much in excess of the
specified depth.

When all the trenches have been dug, the pegging and levelling
is undertaken. For this operation a number of pegs (either wood
or odd pieces of conduit, gas-pipe, etc.), a builder's spirit-level
and a levelling board are used. The latter item is a board about
10 ft. long, 6 in. deep and an inch thick, that has been planed
true on the two opposite edges.

Starting at the lowest corner, a peg is driven in in the centre of the trench until it is the required 9 in. depth above the subsoil. Further pegs are driven in at intervals of several feet, and, in any case, close enough together to use the levelling board on them. The pegs are each levelled from the first peg, using the spirit-level and board, regardless of the height of the peg above the trench bottom. During the levelling operation the spirit-level and board together should be turned end for end as each subsequent peg is driven in. This will compensate for any possible inaccuracies in the levelling board.

When a complete length of trench has been levelled in this manner, each peg is checked for height. Any instances where the depth is under the 9 in. required for the concrete are rectified by cutting the trench bottoms to the correct depth in the affected places.

In the case of stepped foundations, the steps are made to equal a definite number of courses of bricks. For instance, with a $2\frac{5}{8}$ in. bricks and $\frac{3}{8}$ in. joints, a step of 6 in. will mean a jump of two courses, a step of 9 in. three courses and so on.

The first peg for the new level is knocked in immediately behind the last peg of the previous level, and the correct step is easily obtained by direct measurement on the pegs.

When the trenches are completed and pegged they have to pass the scrutiny of the Building Inspector before the concrete is placed in position.

Whilst the foregoing has been taking place, the materials for concreting (cement and gravel) should have arrived on site, and as soon as the trenches have passed inspection this stage of the work can proceed right away. Shovels and a barrow will already be on the site, and now a few short planks will be required. These enable the concrete to be barrowed over the outside trenches to reach those for the internal walls.

The concreting of the foundations is a job which the reader should certainly be able to tackle on his own. With a couple of helpers the foundations of an average house can be put in during a weekend, using a hired mixer.

If casual labourers are employed, they can either supplement one's own work, in which case it is normal to pay by the hour, or, if the reader does not feel like taking a hand in the operation, it is more usual to obtain a price for the completed job.

When employing casual labour, and also contractors doing complete jobs, it is advisable to check the insurance position before employing them. Broadly speaking, contractors doing such work as roofing and other major jobs should be covered by their own insurance, and the person employing them may only need to cover himself for Public Liability.

In the case of casual labour, it may be necessary to cover for both Public and Employer's Liability.

To cover for both can be quite reasonable, but before any work is started on the building it would be a good idea to have a word with one's own Insurance Agent, or to visit the local office and obtain first-hand information.

When hiring a mixer, consideration should be given for the number of people engaged on concreting. The larger the gang, the larger the capacity of the mixer that can be used, and the labour can be adjusted to enable the mixer to run continuously. Two people can be continuously filling the mixer and the rest wheeling the barrows.

It is also a good idea to have somebody in the trench directing the tipping operations, and levelling off the concrete as it is placed. This is done with the back of a spade until the concrete is level with the tops of the pegs. The concrete should be compact, but undue floating should be avoided.

When placing the concrete, care should be taken that the barrows are not tipped too near the edge of the trench, otherwise the sides will collapse under the weight.

Finally, concreting should never be undertaken during frosty weather, and if there is possibility of rain the trenches should be covered up until the concrete has gone off sufficiently to be unaffected.

Specification clauses

This stage of the work is covered by the following clauses in the specification. Nos. 1, 3 to 5 inclusive, and 7 to 15 inclusive.

Clause 3 is more likely to refer to escavations for drains rather than foundations. All the other items are self-explanatory.

Quantities of materials

The only materials required for the foundations are cement and aggregate, and for this instance it will be assumed that the latter is a well-graded 'as dug,' all in aggregate. This means that the coarse stuff and the fines, or the sand, are mixed together. A mix of 1 : 8 is also assumed.

The first step is to check the volume of the concrete required in the foundations.

For 11 in. walls the cross-sectional area of the foundation is the 2 ft. width multiplied by the 9 in. depth, or 2 ft. × ·75 sq. ft. which equals 1·5 sq. ft. The average house discussed in this book will have strip foundations for its 11 in. walls of approximately 100 ft. length. The volume of the finished concrete will therefore be 100 ft. × 1·5 sq. ft. which gives 150 cu. ft.

For the 4½ in. walls the cross sectional area of foundations will be the 1 ft. 6 in. width multiplied by the 9 in. depth, or 1·5 ft. × ·75 ft. which equals 1·125 sq. ft.

The total length of these foundations will be in the order of 44 ft. The volume of the finished concrete will therefore be 44 ft. × 1·125 sq. ft. which gives 50 cu. ft.

The total finished concrete required is 150 cu. ft. for the 11 in. foundations plus the 50 cu. ft. for the 4½ in. foundations, which gives a total of 200 cu. ft. Now although 200 cu. ft. of finished concrete is needed, more than 200 cu. ft. of cement and aggregate will be required. This is because when the materials are mixed together with water, the cement and the sand of the aggregate fill the interstices of the coarse material with a corresponding reduction of the total volume.

If the figure of 200 cu. ft. is divided by ⅔ a figure of 300 cu. ft. is arrived at, which represents the total volume of both cement and aggregate.

As previously stated, a mix of 1 : 8 has been assumed, and this splits the 300 cu. ft. total into 33·3 cu. ft. of cement and 266·7

cu. ft. of gravel. The 33·3 cu. ft. represents a ninth portion of the
total, and the 266·7 cu. ft. represents $\frac{8}{9}$ths, thus giving the proportions of 1 : 8. Now a hundredweight bag of cement is equivalent to
1¼ cu. ft., and so the number of bags needed for the foundations
is 33·3 cu. ft. of cement required, divided by 1¼ (or 1·25) which
is 27 bags, or more nearly 28.

As far as the gravel is concerned, a cubic yard is equal to 27
cu. ft. and therefore the amount required in this instance is 266·7
cu. ft. divided by 27 cu. ft. which gives 10 cu yd.

When ordering these materials it would be advisable to increase
the order for the cement to 30 cwt. or even 40 cwt., and the gravel
up to 12 or even 15 cu. yd. This gives a price advantage as cement
is cheaper by the ton, half ton or even by 5 cwt. lots, whilst gravel
is usually delivered in 4 or 5 cu. yd. lorries, and a split load has a
corresponding price increase per yard.

In any case it will not hurt to slightly over order on these first
two items as any gravel left over can be used on the solid floors,
and the cement will be used by the bricklayers following on as
soon as the foundations are completed.

The quantity of foundation material required for a bungalow
will obviously be more than for a house, but the method of working
out the quantity is exactly the same.

Costs

In discussing costs it is very difficult to be specific, as both labour
and materials vary in different parts of the country. The type and
complexity of the proposed dwelling will also have a bearing on
the cost.

The prices quoted in this and subsequent chapters are for a
small house of 1000 sq. ft. area or under, and of reasonably simple
construction, and in any case, they are only intended as a guide.

Prices are only quoted for the work done in each chapter, but
obviously some of the materials will be bought in bulk, and their
use may extend to several sections of the building. For example,
in Chapter 4 the price is given for the brickwork from foundations
to D.P.C. but unless special bricks were being used, the ones for

this portion of the work would be part of the bulk brick order.

As far as this chapter is concerned, the costs are quite low, the materials being only cement and gravel, and the labour, if any, is employed just for marking out, digging and concreting. The probable allocation of costs would be as follows:

Average summary

	COSTS	
	As built	Using own labour
	£ P	£ P
Marking-out	10·00	—
Digging and concreting	35·00	—
Hire of mixer, plus haulage	5·00	5·00
Cartage of surplus material	15·00	15·00
28 cwt. of cement	15·40	15·40
10 cu. yd. 'as dug' gravel	30·00	30·00
Total for this stage	£110·40	£65·40

Four

Brickwork up to Damp Course—Concrete sub-floors

Building Regulation requirements

The Building Regulations which affect the next stage of building are those which apply to the cavity walls generally, damp-proof course and site concrete.

For cavity walls, the requirements are that the wall shall be of sufficient thickness to ensure stability of the structure as a whole, the cavity shall not be less than 2 in. or more than 3 in., and that the brickwork be properly bonded and the leaves tied together with metal wall-ties.

All external walls of a domestic building must resist the penetration of rain or snow, and also resist the passage of rising damp, the latter item being taken care of by the inclusion of the damp-proof course. This must be at least 6 in. above ground level. The cavity in these walls must also extend to at least 6 in. below the D.P.C. That is, any infilling of the lower portions of the cavity wall to prevent moisture lying there, must not extend up from the foundations as far as the D.P.C.

The majority of these items, and those covered in other chapters, are normally taken care of by the usual good building practice, but it does help to have a knowledge of the requirements.

The regulations concerning the sub-floor concrete are that the turf and vegetable matter shall be completely cleared from the area covered by the building, and that the concrete, properly laid

c

on hardcore, shall have a minimum depth of 4 in. The mix is also specified as being a maximum of 1 : 9.

There are also requirements as to the fire resistance of the external walls, but these are automatically observed with the normal 11 in. cavity construction.

Information and detail on plans

Once again it is the section which will show and detail the items discussed in the chapter.

The ground line must be indicated, and both the internal and external walls will be shown extending upwards from the foundations to 6 in. above the ground level.

The D.P.C. will be indicated by a heavy line across the walls, remembering that there is a separate portion of D.P.C. for each of the leaves, or skins, of the cavity walls. The D.P.C. is not continuous across the full 11 ins.

The top of the sub-floor concrete is then shown level with the damp-proof course, the thickness extending below the level for 4 in. The concrete is sectioned in the standard method.

The hardcore is shown below this, and the sectioning is by diagonal lines in alternate directions.

Constructional details

Having successfully completed the foundations, the next stage consists of bringing the brickwork of the house up to the damp-proof course, filling with hardcore and laying the sub-floor concrete.

Bricklaying is one of the jobs which the reader will, quite rightly, have very definite views about his capabilities of tackling. Without experience of the trade this is a part of house building which will be sub-contracted, and the bricklayers should have already been contacted so that they can start work as soon as the foundations are ready for them.

There is, however, one method by which the reader could take an active part in this stage of construction. That is by acting as a labourer for either one or two bricklayers who were working on the house in their spare time.

This was, in fact, the method used by the author in building a pair of semi-detached houses. The brickwork for these houses was done at weekends, during the summer evenings and holiday times. Even one week's summer holiday was spent in this manner. By acting as a labourer a slight reduction was obtained in the price per thousand for laying the bricks.

This method has the disadvantage in that it is slower than using full-time bricklayers, and the reader may feel that the slight saving may be more than offset by the extra time taken.

A lot depends on the individual circumstances. If unlimited time is available, savings like these, however slight, make an appreciable difference when the house or bungalow is completed.

Brickwork up to damp course, or D.P.C. as it is known, is built in common brickwork until about three courses below the actual level of the damp course. The damp course is always 6 in., or two courses, above finished ground level. By changing from common to facing bricks at the point mentioned, it ensures that the finished appearance of the house is preserved.

If the site is sloping, and it is not intended to bank up to the level of the various entrances, the change from common to facing bricks should be made as the brickwork emerges from the ground.

Sometimes brickwork used for the foundations is done in semi-engineering bricks. These bricks are slightly cheaper than the majority of common bricks and are of a very hard texture, being midway between an ordinary common brick and an engineering blue brick. They are not of such a regular shape as a good common brick, and, therefore, from the bricklayer's point of view, not so easy to lay. As well as being used for foundation work, this type of brick is also used for building up manholes and work of that nature.

If it is possible to obtain a supply of this type of brick locally, and you decide to use them, it would be advisable to obtain the views of the Local Authority as to their suitability.

One other point that should be mentioned before starting the brickwork is the exact position of entry for the gas, water and electricity services. The exact requirements for these are discussed in a later chapter, but the consideration of their position is necessary at this stage to save subsequent digging and knocking out brick-work.

The best method of entry for these services is *via* a short length of salt-glazed drain-pipe laid across the cavity at the points of entry. Generally speaking, the water will be the lower of the services as there is a minimum distance below ground required by the Water Authority. Hence, it is quite often run in the foundation trench, and if this is the case the point of entry can be immediately above the concrete footings. The point of entry for the other two services is likely to be two or three courses below ground level.

Once the foregoing points have been settled, brickwork can start on the bottoms.

Before laying any bricks, small stacks are positioned round the edges of the trenches for a working storage, and spot-boards placed between them on the side of the house that it is intended to work. The cement mortar will be mixed on a larger spot-board adjacent to the store of sand, and, after mixing up, one or two bucketfuls are transferred to the small spot-boards around the trench. These spot-boards, plus the piles of bricks, ensure that the bricklayers have continuity of working materials along the length of the trench. The same applies throughout the construction of the house, although keeping the bricklayers supplied with materials becomes more arduous as the work progresses and scaffolding is erected.

To start the brickwork, the profile boards come into use again. Lines are stretched across saw cuts denoting the outer skin on two adjacent sides. Their intersection marks the corner of the house, and the bricklayer then plumbs down from this position and marks it on the foundation concrete. If two bricklayers are working on the job, the second one will do exactly the same thing on an opposite corner.

Once these positions have been marked they then proceed to set up these corners in brickwork.

The method of doing this is to lay several bricks of the first course in both directions, and then 'racking back' in order to get the correct bond. Having set up corners, a pair of bricklayers will then proceed to run in the courses, until the brickwork finishes level and then the whole procedure starts again.

The height to which this brickwork rises to D.P.C. level may be 2 ft. 6 in. to 3 ft., depending on the nature of the subsoil and whether the site is sloping or not. This height represents ten or twelve courses of brickwork. As this work proceeds the two leaves of the wall must be tied together with wall-ties.

There are two types of wall-ties. One is a flat galvanised strip, split at each end to build into the brickwork, and with a twist in the centre to prevent moisture in the cavity running back to the inner skin. The other type of wall-tie is made from galvanised wire, and once again there is a twisted portion in the centre which prevents the passage of moisture to the inner skin.

These ties should be built in every fifth course and placed not more than 3 ft. apart.

When the brickwork reaches D.P.C. level, the cavity is filled with weak concrete to within not less than 6 in. of the finished level. This concrete infilling helps to strengthen the wall and also prevents moisture from the ground penetrating the outer skin of brickwork and then lying at the bottom of the cavity.

The brickwork should be left down one course on the outside skin at those positions where it is intended to have an outside door. Doing this facilitates the later construction of the step.

Once the brickwork has reached D.P.C., the bricklayers keep off the job until the sub-floor concrete is put down.

This next stage is one that can be done by the reader if he is aiming at doing as much work as possible himself.

With the brickwork at D.P.C. level, the insides of the rooms will have the appearance of 'islands,' there being a gap all the way round between the upstanding soil and the brickwork, due to the latter being set in on the foundation concrete. If the vegetable soil has not been cleared over the whole site, this job will have to be done before any hardcore is deposited. Barrowing away vegetable soil at this stage is a very awkward job, as it has to be removed over the surrounding brickwork. It can be done this way, but obviously it is far easier to clear the top soil from the area occupied by the house beforehand.

Another job that has to be done before the hardcore is added is to backfill the small trench left between the soil and the inside face of the inside skin. This is done by using the spoil from the trench excavations, and carefully filling this 6 in. space and ramming down in layers with a punner.

One other job that must be done at this stage is to provide ducts to allow the services to come up through the hardcore and site concrete. The best way of doing this is to build rough brick ducts, three-sided, the fourth being the inside skin of the outside wall. As mentioned previously, the one for the rising main will probably be the deepest, and the bricks can be laid with cement mortar. If there is any depth to the foundations where the water enters, the duct should be formed large enough to manoeuvre the rising main through the wall from the outside and up to site concrete level.

The entry for the other services are generally higher from the foundation concrete, being normally two or three courses below D.P.C. The ducts for these can be built dry, that is, the bricks laid on top of one another without cement mortar. They are laid on the hardcore when it has been built up to the level of the particular entry.

Hardcore is the next requirement, and the actual filling will depend on supplies available in the locality. Bricks make an ideal filling and in some areas there are other materials available. Whatever material is used it is important that it is laid carefully, so that there will be no settlement once the sub-floor is laid, due either to overlarge pieces of hardcore which do not bed properly and are liable to move, or to voids which are liable to collapse.

One should beware of friends who happen to have 'a load or two of hardcore for the fetching.' Quite often this material is uselss for the purpose. The time and trouble spent on organising transport and labour would have been better spent on obtaining direct supplies from a reliable source.

The minimum depth for sub-floor concrete is 4 in., and the hardcore filling should be stopped level at about 5 in. from the surrounding brickwork. This allows for a layer of 'blinding' all over the hardcore. This blinding consists of ashes or gravel, and its purpose is to give a bed for the site concrete and to ensure that a proportion of it is not lost in the gaps of the hardcore.

Care should be taken that the hardcore and blinding are level. If in any of the rooms it drops in any place, extra concrete will be required, and if it rises up in any place there will be less than the stipulated 4 in. minimum. The concrete should be thickened up to 5 in. where any hearths occur.

The levels can be checked by putting a scaffold board across both the blinding itself and the walls of the rooms.

Once this stage has been reached the Building Inspector will visit the site before the laying of the concrete.

As soon as the site has been inspected work can commence. Laying site concrete is certainly a job that can be carried out by the owner, providing some help is available.

It is necessary to protect floors constructed next to the ground from rising damp, and there are several ways in which this may be achieved. One of the most common methods is to use a PVC membrane which is placed in position before the sub-floor concrete is laid. This method has the advantage that, when the concrete is placed, the membrane prevents the water soaking away to the blinding and the hardcore.

If no membrane is to be used at this stage, and it is decided to incorporate a moisture barrier between the sub-floor and the floor screed (as discussed in a later chapter), then the blinding should be thoroughly wetted before any concrete is placed.

The concrete mix should be to the specification, and the best policy is to lay a room at a time, starting nearest the point of mixing the concrete. Thus when the concrete of this section has gone off it is far easier to wheel barrows across it to the next section. If two or three sections are to be laid immediately after one another, obviously this point does not arise, but otherwise it is a point to consider. In any case, a certain number of scaffold planks will be needed to take the barrows over the hardcore.

The concrete is placed across the section and after a full strip has been placed, say about 3 ft. wide, the concrete is tamped level. This is done by using a scaffold board on edge and guiding it from the brickwork at D.P.C. level. The surface of the concrete is left as a tamped finish. As the strips of the concrete are finished tamped the scaffold board should be retained on edge by placing a brick either side of it at each end, and on the brickwork or site concrete on which it rests.

A further point is that at the threshold of doorways between the various rooms the brickwork will have been left down a course, and the concrete continues right through. If only one section is being done at a time, then a piece of shuttering will be required at any doorway that may occur.

Before leaving the subject of sub-floor concrete, there is an innovation which is worthy of mention. This is the power float, and by its introduction and use the laying of the screed for the floor tiles (discussed in a later chapter) can be eliminated. This not only saves money, but has the additional advantages of greater compaction and strength, great hardness, and the normal drying time for a separate screed is done away with.

The power float is similar in construction to a rotary floor polisher, but instead of the brush it has a rotating disc 2 ft. in diameter. Complete with its $3\frac{1}{2}$ h.p. petrol engine it weighs 280 lb., this weight, plus the engine vibration, giving the great compaction to the concrete.

When a power float is being used, the concrete is placed on the hardcore in the usual manner, but instead of being tamped from the brickwork at D.P.C. level, $\frac{3}{8}$ in. mild steel flats are placed on the tops of the bricks, and the concrete is tamped from these. These strips are used to take out any unevenness in the brickwork, and they also allow the float to work clear of any obstructions.

The float is worked quickly over the area to search out any weak spots, which can be rectified by filling and retamping. When this is satisfactory the floor is then finished floated. The final compaction is then to a density which will allow light traffic almost immediately.

Both large and small building contractors are now using this method of laying sub-floor concrete, and it would be well worth inquiring if there is a power float in the vicinity. If there is, there is no reason why the contractor should not be approached as to its use on a single oversite.

As soon as the concrete is laid, the damp-proof course can be placed in position on all the walls, including the jambs of any fireplaces. The D.P.C. is simply unrolled, the tissue paper removed, and the material laid in strips along the walls. Where one piece of D.P.C. ends and another begins, the overlap should be about 4 in. Bricks should be placed at strategic intervals to stop the strips from blowing off the walls.

When this work is completed a further card is sent in to the Building Inspector who examines the work to date before further building can proceed.

Whilst waiting for the Building Inspector the time can be spent filling in the gap between the external walls and the sides of the foundation trenches.

Specification clauses

The trades in the specification covering this section of the work are Excavator, Concretor and Bricklayer. Under Excavator, Clauses 5 and 6 are applicable. Under Concretor the clauses concerned are 7, 9 to 14 inclusive, and 16 and 17.

Under Bricklayer the following clauses apply to this stage of the work, 35 to 44 inclusive, 46 to 48 inclusive and 61.

Quantities of materials

To work out the quantities of bricks required, it is necessary to know that for the normal stretcher bond there are forty-eight bricks to the square yard of brickwork, for a single 4½ in. thickness of wall. A panel of brickwork exactly a yard square will be twelve courses of 2⅝ in. bricks high, including the joint, and there are four bricks in each course. Hence the figure of forty-eight. For 11 in. cavity work this figure is obviously doubled.

Assuming 100 ft. of 11 in. wall, this will be 200 ft. approx. of 4½ in. work, or 67 yd. At 3 ft. from foundations to D.P.C. this is then 67 sq. yd., and the quantity of bricks required is 67 × 48, which is approx. 3200.

If the length of the internal walls is taken as 44 ft. this is 15 yd. approx. which means 15 × 48 bricks, which is approx. 800. The total is therefore 4000 bricks.

In actual fact about 300 bricks out of this total will be facing bricks where the cavity walls emerge from the ground. However, to allow for wastage it would be a fair assumption to say 4000 commons and 400 facings for the dimensions of walls quoted.

The method of working out the quantity of cement and sand required for the cement mortar is similar to that of working out quantities of concrete. Once again it should be noted that these materials will not be delivered in these small amounts, and these figures are given in order to show how both the quantities and the costs are proportioned, and also to give the reader a guide in order that he may be able to make similar calculations for his own house or bungalow.

The total brickwork measured in $4\frac{1}{2}$ in. work is 67 sq. yd., plus 15 sq. yd., giving a total of 82 sq. yd. It takes ·024 cu. yd. of cement mortar (1 : 4) per sq. yd. of brickwork, so the amount for 82 sq. yd. is ·024 multiplied by 82, which is approx. 2 cu. yd. or 54 cu. ft.

In this case it is assumed that the yield is the same as the total volume of the sand.

Therefore, with a 1 : 4 mix, 54 cu. ft. of sand will require 13 cu. ft. of cement.

The amount of cement is therefore 13 cu. ft. divided by the $1\frac{1}{4}$ cu. ft. volume of a hundredweight bag of cement, which is ten bags.

The sand required is the 54 cu. ft. divided by the 27 cu. ft. volume of a cubic yard of this material, and this gives 2 yd.

The amount of hardcore required can simply be obtained by working out the volume to be filled in each of the sections of the brickwork. If the figure in cubic feet is divided by twenty-seven, it gives the number of cubic yards required.

An average figure could be in the region of 18 cu. yd. of hardcore for an average depth of a foot, and 9 yd. for an average depth of 6 in.

The method of working out the amount of site concrete required is exactly the same as working out the foundation concrete explained in Chapter 3.

Five hundred square feet of sub-floor concrete at 4 in., or a third of a foot deep, is 165 cu. ft. of concrete. To give this actual

yield 250 cu. ft. of cement and gravel is required, obtained by
dividing the 165 cu. ft. by the constant two-thirds.

With a mix of 1:6, 250 cu. ft. is approximately proportioned
35 cu. ft. of cement, to 215 cu. ft. of gravel. With cement at 1¼
cu. ft. per cwt., and gravel at 27 cu. ft. per yard, this is 28 cwt. of
cement to 8 yd. of gravel.

In exactly the same way, at a 1 : 12 mix, the infilling concrete for
the cavity of the average size of house will be composed of 3–4
cwt. of cement and 2 to 2½ yd. of gravel. Figures for bungalows
would obviously have to be adjusted accordingly.

Finally, the only other material which needs a quantity check
for ordering purposes is the D.P.C. This is usually sold in rolls
8 yd. long, so it is a fairly simple matter to add up the lengths of
walls, not forgetting that the 11 in. walls need a D.P.C. on each
skin. The quantity required at this stage will be about ten or
eleven, not forgetting that a roll of 9 in. D.P.C. is required for the
chimney jambs.

Costs

The main new item in this chapter as far as costs are concerned is
the brickwork, both in the costs of the bricks and their laying.
Costs of the bricks can vary greatly, but as an average, £14·00 per
thousand would be reasonable for facings, and £9·70 per thousand
for commons. These can either be obtained from a local brick-
works if there is one in the vicinity, or from any large company
which distributes in the area.

As far as laying the bricks is concerned, £15·00 per thousand is
an average figure and this should include the erection of scaffolding, fixing door- and window-frames and the shuttering and
casting of lintels.

If the reader is prepared to act as a bricklayer's labourer, it is
possible that someone may be found who will be prepared to take
on the job at weekends and light evenings for a lower figure, say
£12·00 per thousand.

The other main point where the reader could save at this stage,
is in placing the hardcore and doing the concrete himself, as this

item can cost £30·00 for a house and up to double this figure for a
bungalow.

Only solid floors have been considered, as these are cheaper than
wooden-boarded floors. Also they afford more scope for the reader
to use his own labour in their construction.

If it is possible to obtain the use of a power float as discussed
earlier in the chapter, a further saving will be made, due to the
elimination of the separate sand and cement screed at a later stage.

The other items such as D.P.C., hardcore and PVC membrane
are all more or less standard prices and there is little other saving
that can be affected at this stage.

Average summary

COSTS

	As built £ p	Using own labour £ p
4000 commons @ £9·70 per 1000	38·80	38·80
300 facings @ £14·00 per 1000	4·20	4·20
Cost of laying @ £15·00 per 1000	64·50	—
Cost of laying @ £12·00 per 1000	—	51·60
Cement for bricklayers—10 cwt.	5·50	5·50
Sand for bricklayers—2 yd.	4·00	4·00
Infilling concrete cement—3 cwt.	1·65	1·65
Infilling concrete gravel—2 yd.	6·00	6·00
Labour for hardcore and concreting	30·00	—
Cement for sub-floor—24 cwt.	13·20	13·20
Gravel for sub-floor—7 yd.	21·00	21·00
Hardcore—18 yd.	13·68	13·68
PVC membrane	5·00	5·00
D.P.C., 4½ in., 11 rolls	3·96	3·96
D.P.C., 9 in.	·71	·71
Hire of mixer	5·00	5·00
Butterfly wall-ties	7·35	7·35
Total for this stage	£224·55	£181·65

Five

Brickwork from D.P.C. to first floor joists—constructional details—door and window fixings—lintels—flue construction—fixing the joists

Building Regulation requirements

In taking the brickwork up from the D.P.C. to the level of the first-floor joists, the general requirements for cavity wall construction will be the same as those indicated in Chapter 4. In addition, there will now be the requirement that whenever the cavity is bridged (as at door and window openings) a D.P.C. or flashing must be inserted to direct moisture away from the inside skin of the wall.

There are also requirements as to the fire resistance of the walls, but these are satisfied by the normal traditional methods of building.

As it can be appreciated, there are several regulations relating to chimneys, flues and hearths, as the efficient construction of these is vital to safety.

The heat producing units themselves, including cookers, are classified as Class 1 appliances or as Class 2 appliances. The former includes solid fuel or oil-burning units having an output rating not exceeding 150,000 British Thermal Units per hour, while Class 2 appliances include gas burning units with an input rating not exceeding 150,000 British Thermal Units per hour.

The regulations prohibit the installation of any heating or cooking appliance discharging the products of combustion into the atmosphere unless it is designed to burn either gas, coke or anthracite. And, although oil is not specifically mentioned as a fuel, the fact that it may be permitted is implied in the section of the regulations which allows the installation of appliances in accordance with the Clean Air Act 1956. Electrical appliances are not covered by the Building Regulations.

It is in relation to these requirements that the regulations for chimneys, flues, etc. have been formulated.

For the basic type of domestic dwelling described here, solid fuel grates and appliances and gas and electric heating units are the most likely means of heating to be chosen, and the constructional details most likely to be involved at this stage are those relating to fireplace jambs and the flues leading from them.

Materials must be non-combustible and the jamb on either side of the opening must not be less than 8 in. thick, and there are also requirements for the thickness of the backs of the openings, depending upon whether they are on an external wall, back-to-back or separate entities.

Chimneys which are built to serve Class 1 appliances must be lined with an approved type of flue liner (or else constructed from approved concrete flue blocks) and, if rebated or socketed units are used the flue must be built with the socket uppermost.

The size of the flue in a chimney serving a Class 1 appliance must be capable of containing a circle of not less than 7 in. diameter, although the flue may be restricted to form a throat. Flues within the same chimney structure must be separated by solid material not less than 4 in. thick. Chimneys for Class 2 appliances must also be built with approved flue liners or, if constructed from dense concrete blocks, these must have inside faces of high alumina cement and the blocks must be jointed with this material also.

There are two other important regulations, and these affect not so much the construction of the house but the design of it.

The first concerns the ventilation of habitable rooms. All habitable rooms should have a window that opens to the external air, and this opening must have an area of not less than a twentieth of the floor area of the room. The area can be made up of several opening portions of the whole window. In addition, part of that opening area must be at least 5 ft. 9 in. from the floor.

Larders, while not classed as habitable rooms, must be ventilated to the open air by means of either one or more windows or by two or more ventilators (capable of being closed and having an unrestricted open area of not less than 7 square inches) which must be built into the upper and lower portions of the larder. If a window is used it must have at least 130 sq. in. capable of being opened, and windows or ventilators must be fitted with durable, fly-proof screens.

The other regulation concerns the height of the habitable rooms, and these should not be less than 7 ft. 6 in. high.

Finally, it is a requirement that the walls, floor and roof shall be constructed to resist the passage of heat from the inside of the house to the outside. Once again, this is an item which is normally covered by standard building practice.

Information and detail on plans

The section and the ground-floor plan both show the majority of the details of this next stage of construction. Sizes and positions of the rooms will be shown on the ground-floor plan, and this will also show door and window positions.

The screened larder window should be indicated, it being necessary to show such items as this as the drawings are checked for Building Regulation approval as well as for planning permission.

Cavity walls are actually shown on the plan and should be indicated as such. The thickness of internal walls should be quoted and the materials of their construction, whether block or brick.

The rooms should all be nominated on the plan together with dimensions showing length and width. There should also be sufficient dimensions to set out door and window positions. There should be overall outside dimensions given, and these should be checked as soon as corners are set up on the foundations.

Room heights are an important point and these are given on the section together with the sizes of floor joists which are usually 7 in. × 2 in. and 7 in. × 3 in. Room heights can be down to 7 ft. 6 in., but this tends to be rather low. A compromise is to have the living rooms at 8 ft. and the bedrooms, if a house is being considered, at 7 ft. 6 in.

Once all this information has been drawn on the plan, and heights of rooms and windows indicated on the section, it is then a fairly easy matter to draw the purely pictorial representation of the various elevations.

Any entrance-porch details, such as projecting canopies and supporting piers, will also show on the elevations.

Constructional details

Once the damp course has been inspected, the next stage is to bring the brickwork up to the level that is required for the floor joists to be placed in position. As the brickwork rises to this level, so door, window-frames and chimneys are built into position.

Window- and door-frames will have already been ordered, and must be on site before this stage of the building can proceed.

Door-frames will be the back and front doors, usually ordered as 4 in. × 3 in. sections, rebated and rounded, with horns for building in. This means that the section is rounded on the edge facing outside and rebated to take the door. The horns are projections at the top of the frame which are built into the brickwork to give stability against the action of the door.

When the frame is delivered to the site, it will have two pieces of wood nailed to it. One will be across the bottom of the frame and the other diagonally across one of the top corners. This ensures that the frame is square and on no account should these be removed until the frame is securely bricked up.

These door-frames are erected as soon as any brickwork is commenced. This is done by knocking a 4 in. nail into a scaffold plank, and leaning the latter against the frame with the nails hooked over the top. The other end of the plank rests on the floor concrete, the end weighted down by a pile of bricks placed over and round it. Before being erected in this manner frames should be given a coat of lead-based priming paint.

The positioning of these frames is done by direct measurement, both along the brickwork and across the cavity. The latter dimension being 3 in., measured in from the front face of the outside wall.

Before the door-frames are placed in position, galvanised nails are knocked into the underneath of each vertical portion of the frame, the heads protruding about three quarters of an inch. The frame is then packed up until it is the thickness of the sand and cement screed above the floor concrete. The protruding nails will then form a key for the bottom of the frame when the finished screed is laid at a later date.

Before any brickwork is built around the frame it should be checked with a level to ensure that it is perfectly vertical.

There are two other items concerning these main frames, one being their attachment to the brickwork, the other the closing of the cavity.

One method of tying the frames to the brickwork is by inserting $\frac{1}{4}$ in. thick wooden pads at approximately every 18 in. as the bricking-up proceeds. The positions of these pads are pencilled on the frame and holding screws can then be inserted in the correct positions.

A far better method is to obtain some lugs bent at right angles. There are two screw-holes in one face for attaching to the frame, and the other face split to a 'fish tail,' in the same way that a wall-tie is split. The lugs are screwed to the frame as work proceeds, the split end being built into the brickwork.

As far as closing the cavity is concerned this is done by cutting bricks and making a 90 degree return with the inner skin. As this work is carried out, so a strip of $4\frac{1}{2}$ in. D.P.C. is built-in vertically. This forms a water bar between the outside skin of the brickwork and the cavity-closing brickwork.

If metal french windows are being fitted, the foregoing remarks regarding setting up with a scaffold plank and closing the cavity apply equally as much. This time, however, the lugs for building in the brickwork are provided with the french doors. They are screwed to the frame and built-in as work proceeds in exactly the same way.

Whilst the door-frames are being built into position, the general internal and external brickwork will be proceeding. The method for doing this is generally the same as for bringing the brickwork up to D.P.C. The bricklayers will set up corners and then run the intervening bricks in with a line. As the work proceeds all the 11 in. cavity work is tied together with wall-ties and great care must be taken to ensure that the cavity is kept clean. One way of doing this is to use a length of batten with a length of strong cord attached to each end. The batten is left in place until just before a set of wall-ties are placed in position. The batten is withdrawn from the cavity and cleaned of any mortar droppings, the wall-ties are placed in position in the wall, and the batten goes back and rests on top of them. This operation is repeated when the next set of wall-ties are placed in position five courses later.

As the facing work is built up, this brickwork will have to be jointed. There are various methods of doing this, and probably the ones most commonly used on houses are the flushed joint, a keyed joint or a struck weathered joint. The choice of the appropriate joint depends on the type of brick being used, and the overall appearance of the finished work.

The flushed joint and the struck weathered joint are formed with a trowel and the keyed joint is formed with a special tool. As its name implies a flushed joint is where the mortar is finished flush with the face of the brickwork. The struck weathered joint is one in which the top of the joint is set in a little way under the brick above it. The keyed joint is a rounded joint and is often done with a portion of a semi-circular bucket handle. It is often known as bucket-handle joint.

Internally there is one job which can be done at this stage and so save a lot of future trouble. That is that round the internal walls vertical joints should be raked out approximately every 2 ft. 6 in. This raking-out is to allow wooden plugs to be driven in when the time to fix the skirting-board arrives.

Where the internal walls meet the inside skin of the 11 in. work, the latter will be keyed out to take the cross walls if it is the intention to build them up at a later stage. Otherwise a start will be made on the cross wall and it will be racked back as the inner leaf of the cavity wall progresses.

As the brickwork rises, openings will have to be built for the window-frames. Whilst being built-in, the frames are held in position in exactly the same way as the door-frames.

When the window-frames are first set in position (once again obtained by direct measurement) the brickwork of both the inside and outside skins must be left down one course. This is to allow for cills to be built in at a later date. The frames themselves are therefore supported on bricks on edge placed across the cavity at sufficient intervals to support the frames.

A vertical D.P.C. is required as the cavity is closed on either side of the frame, and the frames themselves must also be built into the brickwork. If wooden frames are used the best method of doing this is to use the fish-tail lug again. Another way which is sometimes used is to knock a pair of 4 in. nails into both sides of the frame, leaving a couple of inches protruding. These protruding nail-heads are then built into the brickwork.

Once the brickwork has reached the top of the window-frame the lintel has to be formed to carry the brickwork which goes on above this opening.

There are several ways of treating this feature, and one method is to use a boot lintel. This only shows its facing for the thickness of a course of bricks on the outside of the building, thickening up to standard lintel depth on the inside skin.

In certain rural districts special reconstructed stone may have to be used and the window opening treatment may be a special feature. An example of this is the Label Mould type of lintel in the Cotswold districts.

Another treatment for a small house is to show a soldier course of bricks to the outside with a normal reinforced concrete lintel supporting the brickwork of the inner leaf of the wall. The soldier course itself is a row of bricks set vertically, their bottom and back edges supported on a piece of angle iron of the order of 3 in. × 3 in. × $\frac{3}{8}$ in.

Where window openings of small span are involved, the concrete lintels may be cast separately and lifted into position when they have set, but for large spans the lintels would have to be cast *in situ*.

When separate lintels are being made they can be cast between two scaffold planks, the planks being kept apart by bricks to give the correct thickness. Bricks can also be used on the outside to keep the planks in position. These lintels can be cast directly on the sub-floor concrete, but building paper should be put down first of all to stop them bonding to the surface.

The problem becomes a little more complicated when the lintels are cast *in situ*, and it is essential that the formwork containing the concrete is well constructed and adequately supported, with props approximately every 3 ft. Both separate and *in situ* lintels will also require reinforcing bars.

The concrete mix for the lintels is the third one in the specification, and this time the sand and the coarse aggregate are separate. This makes a better mix structurally and is more suitable for lintels than using the 'all in' type of aggregate.

The concrete should be well consolidated as it is placed, and reinforcing bars are added as soon as the first inch has been placed over the bottom of the shuttering. The bars are spaced equally across the lintel, and the rest of the concrete is then poured and well tamped. Care should be taken that the reinforcing bars are not disturbed during this operation.

For openings with a span up to 4 ft. reinforcing bars should be $\frac{3}{8}$ in. diameter and two in number. For 5 ft. openings two $\frac{1}{2}$ in. bars are sufficient, and from 5 to 8 ft. two $\frac{5}{8}$ in. bars are needed.

After pouring, the shuttering should not be "struck' or taken down for at least six to ten days, depending on whether it is summer or winter.

When brickwork is resumed above the lintel, a piece of $13\frac{1}{2}$ in. D.P.C. is built in over it, and for the full length of the lintel. This piece of D.P.C. is built in under the first course of the outside skin, and over the top of the second course from the lintel on the inside skin. The D.P.C. will thus slope down from the back to the front. This prevents any moisture which may get into the cavity above penetrating through the head of the window.

If fireplace openings are needed these will also have to be dealt with when the brickwork begins to rise above the sub-floor concrete. The nibs or jambs which were built into the foundations still continue up in this form until lintel height is reached, usually about 3 ft. from the concrete.

The lintel can be of the normal concrete variety, but it is far

better to use a preformed refractory throat unit. These obviate
the need for a lintel and are so designed that they insulate the fire
surround from excessive heat and also operate with maximum
efficiency.

If the fireplace opening is on an outside wall, and it has been
decided not to make an internal feature of the chimney breast, then
the opening can be built so that the projection is external.

The jambs are built up to the height where the lintel or throat
unit is placed in position, the flue gathered over by cutting and
corbelling the brickwork to bring the flue to the size where the
flue liners can be incorporated. As the flue progresses these
liners must be properly jointed and pointed, and care must also be
taken to see that the regulations are met if two flues are incorporated in the same chimney structure.

In the case of a flue for a domestic kitchen boiler burning solid
fuel, the flue will usually begin about 5 ft. 6 in. from floor level.
This is because this type of appliance usually has to have a length
of 4 in. diameter flue pipe between the appliance itself and the
commencement of the flue.

If this is the case the flue can usually be constructed on a
purpose-made concrete lintel which can be supported on brick
jambs, or else built into adjacent corner walls. If the chimney
structure is external the flue pipe can be allowed to enter the
wall forming part of that chimney, providing that the necessary
regulations are observed.

Where gas fires are concerned the flues serving them should be
lined with approved lining materials, but in some cases an unlined
flue may be allowed providing that certain stipulations as to
length and size are observed.

As the brickwork rises, internal door openings will have to be
formed in the cross walls. These are fairly simple in their forma-
tion. They are built to contain a standard door lining which is
2 ft. 8 in. overall and the brickwork should be $\frac{1}{2}$ in. larger than
that, so that the lining can be placed in position later. The opening
is topped by a lintel which can be cast on the sub-floor concrete
in the manner already described. A point to remember when
making separate lintels is that, as the concrete is going off, the

word 'top' should be scratched on the top face to ensure that, after handling, the reinforcing bars are always at the bottom. When the lintel is in position above the opening the brickwork continues on above it. Wooden pads should be built into each side of the opening as it progresses for the subsequent attachment of the door lining.

The brickwork continues generally until the correct height is reached (either 7 ft. 6 in. or 8 ft.) for the first-floor joists to be put into position.

In working out the layout of the joists, consideration must be given to their spacing, the clear spans and openings and other places which have to be trimmed. It is here that the tables contained in the Building Regulations come in useful, for these detail the maximum span of joist permissible when related to spacing and size.

If possible, the joists should be spaced at 16 in. centres, as this facilitates the subsequent fastening of the plasterboard to the underside. An average size of joist is 7 in. × 2 in. which at 16 in. spacing has a clear span of 12 ft. There are, however, instances where 6 in. × 2 in. are used. For trimming joists round stair-well openings and round chimney-breasts, the size would be increased to 7 in. × 3 in. or 6 in. × 3 in.

The reader may wonder whether he could manage the installation of the joists himself. For someone who is engaged on the difficult task or building anyway, the answer is probably yes. The most difficult part would be the cutting of the tusk tennons in the trimming joists. If no one can be found who is willing to undertake this part of the job only, it may be possible to act as a mate to a carpenter and so cut the cost that way. In any case if the reader decides to tackle the majority of the work on his own, some labouring help will be needed to get the joists into position across the walls.

Joists are cut to length and laid across the walls at the correct spacing. They are checked for being level and if they are not they should be packed where necessary, to bring them to the correct position. Care should be taken to see that no joist comes nearer than $1\frac{1}{2}$ in. to the rendered face of any chimney, as this is a Building

Regulation requirement. There will be no rendering on the chimney-breast at this stage so no joist should be nearer than 2 in. to any chimney brickwork.

Where it is intended that a clinker-block wall is to be built directly on the first floor, it is the usual practice to bolt two joists together to take this load. These instances occur where a wall is required on the first floor without one on the ground floor directly underneath it to continue on up.

In instances where comparatively short lengths of joists are spanned from a wall to a trimmer or trimming joist, the joint between them is a square housed joint. This is not as complicated as the tusk tenon and consists of a simple notching of the smaller-size timber to the larger. When complete, these joints are spiked with nails. The tusk tenon should be held in place by a wooden peg.

When all the joists are in position, lengths of 2 in. × $\frac{3}{4}$ in. tile battens or similar strips of wood should be nailed across the joists to retain them in position whilst the bricklayers build-in between the ends of them. Before being built-in, the ends of the joists should be treated with one of the various liquids on the market as a protection against wood rot.

Joists may either span across individual rooms or in certain instances they can span right across the house and embrace two rooms. There they are cut to go across individual rooms, the joists spanning the adjoining room should be placed at the side of the joists for the first room, where they ledge on the common wall. On no account should joists be placed end to end as each joist would only have $2\frac{1}{4}$ in. of bearing.

Once all the joists are fixed in position, it only remains for a single row of strutting to be placed between the joists and in the centre of the rooms. This gives rigidity to the joists. Probably the reader is already familiar with one method of doing this, and that is the herring-bone type of strutting, which consists of 2 in. × $1\frac{1}{2}$ in. pieces of timber, positioned diagonally between the joists. Sometimes solid strutting is used, which consists of solid pieces of wood equal to the depth of the joists and placed between them. Every alternate strut is offset so that they can all be nailed into position between the joists. There is one further method of strutting which is known as fillet strutting. This is very similar to the solid strutting but much smaller pieces of wood are used, and they are placed at the bottom of the joists.

Whilst the positioning of the joists is taking place, the bricklayers can return and build in the ends. This consists of building the brickwork between the joists where they rest on the inner skin, and carrying on with the outer skin where it was left down to allow the joists to be fixed.

If any form of wood preservation treatment is required, either for the complete joists or simply the built-in ends, it will have to be done before any building-in takes place. If it is not intended to treat all the timbers, the ends in the brickwork should certainly be treated by dipping in creosote.

Finally there is the question of scaffolding, for this will become necessary as this stage of the building develops.

It is usually possible to hire the complete scaffold for a house on a monthly basis. Scaffolding is usually erected by the bricklayers, and if the reader is acting in a labouring capacity he will obviously assist. He should then, at a later stage, be quite capable of raising a section of the scaffolding a complete lift ready for the bricklayers to continue.

The usual type of scaffold employed in a small dwelling is a putlog scaffold.

This consists of upright members, or standards, just over 4 ft. away from the building, and placed approximately every 6 to 8 ft.

The standards are tied longitudinally with ledgers. These in turn support the putlogs which go back to the building. Putlogs are short pieces of tube which have one end flattened. These ends are wedged into a raked-out joint. The scaffold plank rests on these and their spacing has to be carefully watched. Planks must be placed carefully so that the ends are supported and do not tip.

Putlog scaffolds should go right round the house and they derive support from this feature. In addition they should be tied in securely at intervals. This is done by coupling ledgers to tubes across window openings on the inside. This prevents any tendency for the scaffold to come away from the wall. Where it is not possible to use such openings, the scaffolding should be strutted with raked tubes which are leaning to the building.

When the planks are in position for a particular lift, toe boards and guard rails should be added.

When using the scaffolding care should be taken not to overload it with excessive piles of bricks.

Scaffolding should always be looked at carefully before use, paying particular attention to the fittings and the placing of the boards.

Specification clauses

The clauses under Concretor which affect this stage of the work are 7 to 11 inclusive, 18–20 inclusive, and 24. The section of clause 10 which is applicable is the one referring to the mix for reinforced work.

The main work in this stage of building comes under Bricklayer. Clauses 35 to 47 inclusive are applicable and also 50, 52 to 54 inclusive, 56 to 59 inclusive, and 61.

The other main trade concerned is that of Joiner. The clauses concerned are 63, 79 and 80. If any form of strutting other than the herring-bone type is used, clause 80 would have to be slightly modified.

Quantities of materials

There is no great difficulty in estimating the quantities of materials required for this stage of work.

The main item is the brickwork. As previously stated, the bricks will probably be ordered on a single order, but it is useful to know the proportions of brickwork in each stage.

There is approximately 100 ft. run of 11 in. cavity work. This means that there will be 100 ft. of facing work and 100 ft. of work in common brick.

Taking the facings first of all, the 100 ft. run will be 8 ft. high, which is 800 sq. ft. of brickwork. This is approximately 90 sq. yd.

and at forty-eight to the sq. yd. this represents 4320 facing bricks.

The number of common bricks for the inside skin will be exactly the same, but in addition there will be the extra 50 ft. run of commons for the internal walls. This is 44 sq. yd. of brickwork and represents 2112 bricks. Added to the 4320 commons already obtained this gives a common brick total of 6432 for this stage.

These figures are only a guide, and will obviously be modified where such things as ceiling heights differ, or where blocks are used instead of bricks. Doorways and windows have also been neglected, but it is a good plan to ignore these (unless very big windows are involved) and this gives a percentage extra for waste. The total of both common and facing bricks is some 10,700 or 223 sq. yd.

Using ·024 cu. yd. of mortar to a square yard of brickwork, this means a total of approx. 5½ yd. of mortar.

Assuming that the volume of sand required is the same as the final yield, this means 5½ cu. yd. of sand.

With a 1·4 mix, this means that the correct proportion of cement will be 1·3 cu. yd. or 35 cu. ft. A bag of cement is 1¼ cu. ft., so therefore the number of bags required will be thirty-five divided by one and a quarter, which is twenty-eight bags.

Some extra sand, cement and ballast will also be required for the lintels. The extra gravel which is required for the lintels will probably amount to 2 to 3 yd., and this should be sufficient for this stage and also for the first-floor openings. This gravel for the lintels will not be the 'all in' type, but graded from 1 to ¼ in.

Additional D.P.C. will be required for window and door openings. The amount of 4½ in. material for closing these could probably be in the region of 100 ft. or four rolls, two rolls of 13 in. should be sufficient for the D.P.C. above door and window openings.

The other major item which has to be ordered for this stage is the wood for the joists. The best way of working out the quantity required would be to draw a plan of the ground-floor walls, and draw in the joists required. If this is done to a scale of ½ in. to a foot, it is quite easy to estimate the timber needed.

The average house will require something in the region of 420 ft. of 7 in. × 2 in. and 20 ft. to 30 ft. of 7 in. × 3 in.

Costs

The brickwork costs for this section of the work will be similar to those in Chapter 4, although they will be greater as more brickwork is involved.

The cost of window-frames will vary, depending on whether wood or metal frames are used. Allowing for french doors and windows in metal, the cost for the ground floor will be in the region of £60·00. The french doors would be approximately half of this figure.

The timber for the first-floor joists is another large item so far as costs are concerned. The cost would probably be in the region of £50·00. The labour for their fixing would be about £20·00 and this sum could be saved if it is decided to do one's own fixing.

Scaffolding will have to be considered now, and its use will extend over several stages. The cost is very variable, depending on how much is required and for how long. The weather can have an adverse affect in the latter item.

Average summary

	COSTS	
	As built	*Using own labour*
	£ p	£ p
4320 facing bricks at £14·00 per 1000	60·20	60·20
6432 common bricks at £9·70 per 1000	62·30	62·30
Cost of laying £15·00 per 1000	161·25	—
Cost of laying—£12·00 per 1000	—	129·00
Cement for bricklayers, 28 cwt.	15·40	15·40
Sand for bricklayers, 5½ yd.	11·00	11·00
Metal windows and french doors	60·00	60·00
Front and side door-frames	10·00	10·00

Timber for joists	50·00	50·00
Labour for joists	20·00	—
D.P.C.	3·54	3·54
Sundries for nails, gravel for lintels, etc.	10·00	10·00
Flue liners	7·00	7·00
Total for this stage	£470·69	£418·44
Scaffolding for three months	£50·00	£50·00

Brickwork from joists to wall-plate—roof timbers —roof covering—gutters and down pipes

Building Regulation requirements

The Building Regulation requirements for this chapter are very similar to those for Chapter 5.

Those clauses relating to the general construction of cavity walls, flues and chimney-breasts are exactly the same, as are those concerning the ventilation, open windows and heights of habitable rooms. There are also regulations relating to the fire resistance of the roof, and to its thermal insulation. In the latter instance a number of insulating materials are listed which may be used to meet this particular requirement. In addition, the roof has to be weatherproof and provided with an adequate method of drainage.

There are also Building Regulations applicable to this chapter concerning the height to which chimneys are built above the roof.

When a chimney is built up through the ridge of a roof its top should not be less than 2 ft. above the ridge. If the chimney protrudes through the slope of the roof then its top should not be less than 3 ft. above the roof, measured from the highest point of the chimney's junction with the roof.

As far as the width of a chimney or a group of chimneys is concerned the requirement is that the least width shall not be less than one-sixth of the height of the chimney above the highest point of the chimney's junction with the roof.

Information and details on plans

This next stage of construction will be mainly detailed on the
section and the first-floor plan. The section should show the height
of the first-floor rooms, the heights of the upstairs windows above
the floor level and details of the gutter and fascia board, etc.

The construction of the roof should be fully detailed. Sizes
should be given for the ceiling joists, rafters, purlins, struts and
the ridge. The pitch of the roof should also be indicated and its
proposed covering. The sizes of these roof timbers can be obtained
by reference to the appropriate tables in the schedule at the back
of the Building Regulations.

The first-floor plan should have its various rooms designated
and sizes indicated in exactly the same way as the ground-floor
plan. Once all these details have been completed the information
can easily be transferred to finish the elevations.

Constructional details

In continuing the brickwork of the house from floor joists to completion, much the same details will be met as described in the
previous chapter.

Externally there will be the first-floor window openings to form.
The method of doing this is exactly the same as previously
described, with the omission of a course of bricks for the cill and
the same method of closing the cavity. The detail above the
windows which come under the eaves may differ very slightly
depending on their height and whether the standard type or
Finlock type of gutter is used.

First-floor windows which are in a gable end, and consequently
have brickwork above them, are identical in their construction to
those in Chapter 5.

Walls which continue up from the ground floor proceed on in the
same way, and door openings are formed as previously described
with blocks being built-in as the opening progresses.

When a chimney-breast is built on up, it is usual to reduce its
section so that it takes up less space in the rooms above the ground
floor. This is done when the brickwork is resumed between the

floor joints, and is achieved by building the flue but not taking the side brickwork out to the full width of the breast. It is not always a good idea to cut this brickwork width down to an absolute minimum as the finished result may tend to look a little absurd. A more reasonable approach would be to build 9 in. of brickwork at each side of the flue, giving an overall width of 2 ft. 3 in. Alternatively it may be desired to offset the chimney-breast when it goes through a first-floor room, in order to make an alcove for a built-in cupboard or a similar item. This is done by building the brickwork on one side of the flue only.

If the chimney is being built as an external projection it is also usual to reduce its section. This is done by cutting and taking over the brickwork until the desired width of chimney is obtained. Once again this can be done equally on both sides or offset by taking over all on one side. A feature is usually made of the point where the reduction in section occurs, by capping the cut brick-work with tiles that match the roof.

If a chimney is an external one, all the work will be in facing bricks. If it is internal, then the work will be in common bricks until it emerges from the roof. The stack will then be built to show facing bricks, and will continue in that manner until its termination.

There are various ways of finishing off a chimney-stack, from a concrete cap to various forms of decorative brickwork.

The concrete cap is quite a common method and it is usually cast in position. It is cast with an overhang all the way round which has a drip groove in it. The cap must be cast with a hole in it for the flue-liner or liners to locate in, and this method of finishing requires quite a lot of shuttering.

A simpler method is to terminate in brickwork, and a decorative effect can be obtained by corbelling out one or two courses just before the finish. Dwarf flue-liners can then be set in position and the whole flaunched with cement mortar.

Flaunching means tapering off the cement mortar from part way up the flue-linings down to nothing when it meets the brick-work at the edge of the stack.

When the stack is finished the joints for the stepped and apron flashings should be raked out ready to receive them. In the case of a stack which protrudes through the ridge of the roof, a brick corbel will be built out to support the ridge board.

The front-porch canopy, if one is fitted, although, strictly speaking, coming in the previous stage, has been deliberately left until now. This is because it is usually more convenient to wait until the first-floor joists are in position. When the joists are in position they give a platform from which the canopy can be handled, an operation which may require several people. (*See* Fig. 9.)

When the canopy is in position it must be properly supported until any piers or other supporting features are built in.

As to the canopy itself, probably the simplest way is to have a reinforced concrete one with an artificial stone facing. The rear of the canopy can be made as a boot lintel to be built into the cavity. The top surface should be weathered, that is, to have a slope to throw off water. There should also be a drip groove on the underside near the edges.

When the brickwork recommences above the canopy a strip of $13\frac{1}{2}$ in. D.P.C. must be built in, in exactly the same way as above the normal window lintels.

Of course, depending upon the design of the house or bungalow, it may be that a prominent canopy is not required, particularly if an inset porch is involved. Consequently, there are a number of ways in which the main entrance may be treated, of which the concrete canopy is the most basic, and alternative structures may be built to incorporate a number of other designs and using a number of other building materials.

Brickwork continues generally until wall-plate height is reached. The wall-plate is bedded on the inner skin of the cavity wall and the outer skin is left down two or three courses to clear the bottoms of the rafters.

Probably the neatest and most satisfactory method of construction is to finish with a course of bricks laid as headers and giving a 2 in. setback on the front of the wall. The wall-plate can then be bedded above this. Using this method closes the cavity and the setback gives a ledge for the soffit to rest on.

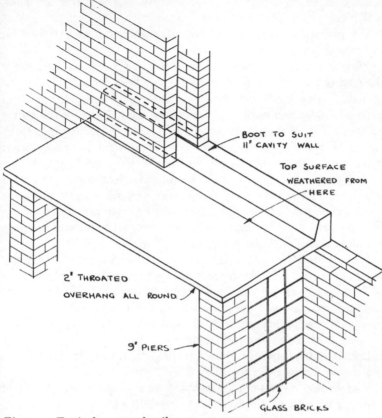

Figure 9 *Typical canopy detail*

If the house has gable ends then the brickwork is continued up to the apex, but the final edge or verge of the gable is not built until the rafters are in position.

At the eaves, the brickwork requires a shoulder or springer, to support the foot of the gable. This serves the purpose of bringing the brickwork out to the line of the roof, where the latter protrudes past the face of the building. Sometimes a small concrete block is used to support the brickwork, or else the bricks themselves can be corbelled out until the desired effect is achieved. Another

D

method is to construct this shoulder in courses of plain tiles, which corbel out like brickwork.

If a Finlock gutter is to be used, then this will have to be built in below any further work is done.

Finlock gutters are precast concrete eaves units, which replace the traditional soffit, fascia and gutters. There are several designs available, and each design has its own stopped ends, internal and external angles and outlets. As the cross section area of the gutter portion is larger than normal, it enables fewer down pipes to be used. The recommended figure is one outlet to 60 ft. run of gutter.

There is also the additional advantage that, due to the double trough construction of the gutter units, the rear one can be used as a lintel. This is done by filling with concrete where the units pass over a window opening. Reinforcing bars are added in the normal manner as in standard lintels. Special lintel supports are available on hire for use when casting lintels.

If the Finlock gutter is being used, the brickwork of the cavity walls is stopped with both skins level. The gutter units are then bedded on top of the walls and is a bricklayer's job. The units are laid level to a line and the joints cut off flush.

The standard block is 9 in. wide and 5 in. and 6 in. closer blocks are also available to eliminate cutting.

Once the blocks are in position they should be lined, and this is a service carried out by Finlock Gutters Limited.

The lining consists of a coat of rubberised pure bitumen, followed by a sheet of aluminium dressed to the gutter contours, the aluminium adhering to the bitumen.

Finally, there is a heavy coat of bitumen with a top dressing of asbestos dust.

Once this stage has been reached the wall-plate is bedded on the back of the gutter units, and normal roofing technique can proceed. An average size for the wall-plate is 4 in. × 2 in. and it is bedded with cement mortar.

Whether the reader would decide to pitch his own roof is a question which only he could answer. A straightforward roof with no hips and valleys and brick gable-ends is a feasible proposition, providing, and this is extremely important, the reader has some knowledge of the part that the various members play in taking the load of the roof, etc.

For a reader who does not feel inclined to tackle the whole job himself, it may again be possible to assist a professional carpenter with the corresponding slight saving on labour charges.

Details of roof timber sizes are set out in the Building Regulations, and the considerations for permitted sizes and spans are the same as for floor joists.

The first stage in roof construction is the placing of the ceiling joists, and spiking or nailing them to the wall-plate. The ceiling joists have to support the weight of anyone who may walk in the roof space above.

If there is any possibility of converting the roof space at any time to rooms, then the ceiling joists should be considered as first-floor joists.

Once again it is advisable to consider the spacing of the joists in relation to the subsequent fixing of the plasterboard. The most adaptable spacing for this will again be 16 in. centres. Reference to the Building Regulations will give permitted sizes and spans for this and other spacings.

Where the ceiling joists are spanning across distances in the order of 12 ft. intermediate support will be required. This is given by a beam which runs across the centre of the span, and by hangers. These beams (usually, but incorrectly called binders) can be built into the brickwork of the gables. The hangers are vertical members from purlins down to this beam.

If the ceiling joists are not continuous across the building, that is, they break on an intermediate wall, it is important that they are well jointed at this point. This will be understood when it is realised that the ceiling joists resist the tendency to splay that the roof has under the weight of the tiles.

The ceiling joists are trimmed round chimney-stacks in the same manner that floor joists are trimmed, although there need be no increase in size, as in the case of the floor joists. An opening about 30 in. square must be formed for a trapdoor to the roof space, and this is done by trimming. When all the ceiling joists are correctly in position, they are spiked or nailed to the wall-plate.

The rafters are the next item to be considered, and the length and size of these depends on the pitch of the roof. The pitch of the roof determines the length of the rafter, and the length of the rafter spanning from wall-plate to purlin and purlin to ridge determines its size.

The pitch is the angle that the rafters make with the horizontal, and this should be given careful consideration.

For concrete interlocking tiles the permissible pitches are between 35 and 45 degrees.

A pitch of 35 degrees uses less timber, but if it is ever intended to use the roof space at a later date, then consideration should be given to head-room. An increased pitch would give greater head-room, but would have a corresponding increase in the timber and roof covering required.

After the ceiling joists are in position the ridge-board is put up so that the rafters can be pitched between it and the wall-plates. The ridge-board will be in the order of 7 in. × 1¼ in. and if it is not obtained in one length it should be jointed with a scarfe joint, the two portions being bolted together.

With the ridge in position a pair of pattern rafters are cut and offered up as a check.

If the roof pitch is 35 degrees, then the angle that the rafter is to be cut to match up with the ridge is 90 — 35 degrees which is 55 degrees.

One end of the rafter is cut on the splay to this angle, and also the bird's mouth. The bird's mouth is a 1 in. × 1 in. V cut in the rafter where it sits over the wall-plate. A guide line is marked on the ridge-board to indicate the point below the top of the ridge at which the rafter should be. The pattern rafter should then fit in any position along the length of the roof.

When the ridge-board and the pattern rafters are correct, the pattern rafters are used to cut the remaining rafters. They are then offered up two at a time (one each side of the ridge-board) and securely spiked with 4 in. nails to the ridge-board, wall-plate and ceiling joists.

When all the rafters are in position the ends of the ceiling joists are cut on the angle to eatch the slope of the rafters.

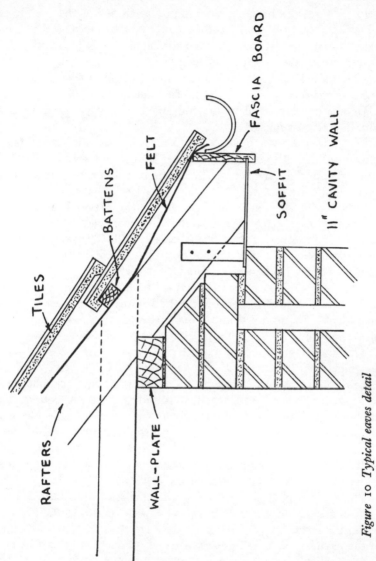

TILES

BATTENS

FELT

FASCIA BOARD

SOFFIT

11" CAVITY WALL

RAFTERS

WALL—PLATE

Figure 10 *Typical eaves detail*

Purlins are the next consideration and these are the members which span along the roof, midway between the ridge and the wall-plate. Their purpose is to prevent the sag of the rafters under the weight of the tiles or snow and wind loads.

Although the purlins are large section members (usually about 8 in. × 3 in.) they do require intermediate support.

Details of permissible sizes and spans are again given in the Building Regulations.

When the purlins have been located in position and finally spiked to the rafters, the ends are built in to the brickwork. Once the purlins are secure the hangers supporting the ceiling joists can be cut and attached.

Support for the purlin is also done at this stage. Usually this consists of two struts (about 4 in. × 3 in.) which oppose one another and rest on a central load-bearing wall. If, by virtue of the construction of the walls, it is not possible to do this other methods of support may have to be adopted.

One method is to continue a section of a load-bearing wall up to and round the purlin. Another method is to put in a deep beam instead of a ceiling joist and bring the supporting strut down to this, restraining the toe by a block nailed to the beam.

The final items to complete the roof timbers are the soffit and the fascia board. The fascia board is usually 6 in. × 1 in. and has a groove in it to receive the soffit. On the fascia board are mounted the gutter brackets and the half-round gutter. Of course, as previously mentioned, if Finlock gutters are used then these items are omitted. To mount the fascia board the ends of the rafters are cut so that they present a vertical face, and the fascia is then nailed to them. To ensure that all the cuts are in the correct place a line is strung along the ends of the rafters.

The soffit, which is the portion which fills in under the eaves, can be strips of asbestos or wood. One edge goes into the groove cut in the fascia board, and the other edge rests on the outer skin of the cavity wall. Small pieces of wood (about 2 in. × 1 in.) are nailed vertically to the rafters at intervals along the length of the soffit. Cross members are added (also about 2 in. × 1 in.), from these vertical members to the toe of the rafters. These soffit bearers then form a member to which the soffit can be attached. If an asbestos soffit is being used, it will have to be pre-drilled before any nails are put in.

Finally, if there is a chimney-stack which protrudes through the roof at any other point than the ridge, it will require some boarding for a back gutter. This is a gutter which is incorporated with the chimney flashing and goes at the back of the stack where it protrudes through the pitch of the roof. (*See* Fig. 11.)

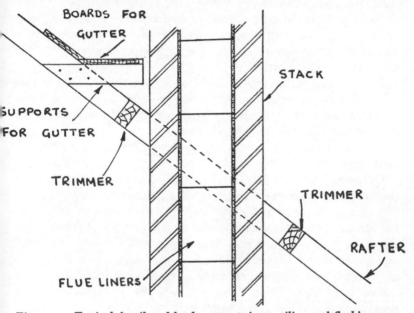

Figure 11 *Typical detail and back-gutter prior to tiling and flashing*

This gutter consists of lengths of $\frac{3}{4}$ in. or 1 in. board. One length is mounted horizontally on lengths of wood attached to the rafters, and the board butts right up to the rafters. A further length is then nailed to the rafters immediately above the horizontal one. A wood fillet is then added where it will come under the tiles. The gutter is then ready for the lead flashing at a later stage.

When all this work has been completed the gutter brackets and the gutter sections are added. Gutters can be either asbestos or PVC and are usually laid with a slight fall, the positioning of the brackets to suit this fall being achieved with a line.

Asbestos half-round gutters have a collar at one end into which fits the plain end of the adjacent section. The gutter sections are drilled and bolted together with galvanised nuts and bolts, the joints being made with a sealing compound. In the case of PVC gutter sections, the sockets have built-in sealing pads and jointing is usually a simple matter of clipping the sections together.

In both types of gutter systems special fittings such as stop ends (which seal gutter ends), returns (which take the gutter round a corner) and outlets (giving a downward connection for the drain pipe) are available.

Down pipes (either asbestos or PVC) are usually added at this stage. These are available in varying lengths and are socketed into each other, the assembly being attached to the wall by means of special brackets.

With all these details complete, the covering of the roof is the next step.

Only single lap concrete interlocking tiles are considered here as they are both economical and of good appearance.

There are several types available but the two most common are the Roman type which have a double roll and the standard type which have a rectangular roll. There are a various range of fittings and special tiles available with the various makes. These should be ascertained before any orders are placed. Sizes of these tiles are in the order of 15 to 16 in. long and between 9 and 13 in. wide, depending on the make and style used.

Once again the reader will have very definite ideas as to whether he will attempt this work or not.

For those who are intent upon doing every item of work themselves, then obviously they will go ahead. Others may feel more like tiling the roof of a bungalow than a house, as it's not so high! Tiling is a straightforward job if the initial setting out is carefully done, and the author knows of a case where the roof of a house was tiled by someone completely unskilled.

The first part is felting and battening the roof. Felt is laid over the rafters and secured by galvanised nails. The felt should have an overlap of 3 in. at the horizontal joints and 6 in. at the vertical joints. The felt should be allowed to sag slightly between the rafters and should also lap over the fascia board. This allows any moisture which may penetrate the roof to drain away to the correct place.

Batten sizes are $1\frac{1}{2}$ in. $\times$ $\frac{3}{4}$ in. or $1\frac{1}{2}$ in. $\times$ 1 in., depending on the spacing of the rafters. Battens are attached by galvanised nails and joints should always be on the centres of the rafters.

Tile manufacturers recommend the gauge or the distance between the battens, at which their tiles should be laid. This can vary between 11 in. to 13 in., giving a head lap on the tiles of 5 in. to 3 in. The head lap is the distance by which one tile overlaps the one below.

The laying of the tiles commences from the eaves, and when the top course of tiles at the ridge is reached it must be a full course. In order to achieve this the gauge of the tiles can be reduced equally over the whole length of the rafters, or the gauge can be reduced on the first two or three courses of tiles immediately above the eaves. The variation in gauge can also be taken up by a number of courses of plain tiles at the eaves.

Assuming that no plain tiles are to be used to affect any alteration in gauge, there are still variations to the treatment of this first course of tiles. This prevents the entry of small birds between the tiles and felt. Some manufacturers make a special block end eaves tile which has the rolls of the tile block solid at the ends. Another method is to lay cement mortar on the felt immediately behind the fascia board to bed the rolls of the tiles. Yet another way is to lay a course of plain tiles along the eaves, so that the tails or bottoms of the tiles comes immediately behind the lower nib of the first course of the main tiles.

Another feature of the roof which needs special treatment as far as the tiles are concerned is the verge. The usual way is to lay an under cloak of plain tiles which project over the face of the brickwork between two and three inches. The actual distance depends on the initial setting out before tiling commences. This under cloak consists of plain tiles which are bedded, sand face down on the brickwork of the gable. The main roof tiles are then bedded to them and the joint struck clean. Any gap between the undercloak tiles and the brickwork should also be pointed. To give a neat appearance to the left-hand verge and to finish on a roll, the flat portion of the tiles affected can be cut back to the edge of the left-hand roll of the tile.

The rest of the tiling is straightforward and the fixing details should be taken from the manufacturers' instructions. It is usual to nail every second course and use one nail-hole. All eaves and ridge courses should be nailed, and also tiles which are adjacent to verges, hips and valleys. Where tiles are against chimney-stacks they are cut as necessary to fit against the brickwork, the final weatherproofing being given by the flashing.

The last tiles to be fixed on a straightforward pitched roof are the half-round ridge tiles. These are laid along the ridge and bedded on mortar. The first and last ridge tiles are slightly tilted as also any which are adjacent to chimney-stacks. This directs water on to the main roof and away from verges and brickwork. Ridge tiles are only bedded completely solid at the joints. At this point a piece of plain tile is laid directly on the ridge to support the cement mortar used for the bedding. Where ridge tiles are exposed on a gable end it is usual to fill the end completely with cement mortar, and set in pieces of plain tile.

The foregoing details of both roof construction and covering apply mainly to a gable-ended building, but the details are very similar if a hipped roof is concerned or even a combination of both.

When all these details are complete the flashing can be done and the roof is then completely weatherproof. The traditional material for this is lead.

The points at which flashing are required are where chimney-stacks emerge from the roof, either at the ridge or through the pitch.

Where a chimney-stack protrudes through the pitch of the roof, the flashing should consist of four distinct sections, two sides, the front and the back. The side flashings are usually in one piece and stepped to follow the pitch of the roof. It is securely wedged into the joints of the brickwork which have previously been raked out, and the free edge is dressed over the tiles about 6 in.

At the back of the stack, a single piece of lead is laid over the back gutter already constructed, one edge being turned up the rear wall of the chimney and the ends being dressed over the tiles. A cover flashing is then fixed to a raked-out horizontal joint and turned down over the upstand or the first piece.

At the front, a lead apron flashing is dressed over the tiles and up the front of the stack, being secured in another raked-out horizontal joint. The bottoms of the side flashings are dressed over the apron flashing.

Where a chimney-stack protrudes through the ridge of a roof, the flashing details are similar to those already described. In addition a lead saddle is provided at the actual junction of the ridge and the brickwork of the stack. After all the flashings are in position and securely wedged the raked-out joints should be pointed up.

Sheet lead for flashings is designated by its weight. For domestic building the recommended weights are 4 to 5 lb. per sq. ft. for side flashings, aprons, etc., and 6 lb. per sq. ft. for saddles.

Although lead is the traditional material for flashings, other materials are now being used. The two most notable are aluminium and zinc. The flashing details for both these materials are similar to those when lead is used. These two materials are designated by gauge and not by their weight per sq. ft.

Finally, to complete this stage of the work, the glazing of the windows is carried out and then the scaffolding can usually be dispensed with.

Glazing is a job which is better done by professional labour, especially where large windows are involved.

Panes of glass should be cut with clearance all the way round to allow for expansion. Corners should also be nipped off to make sure there are no tight fits. Ordinary putty is unsuitable for use with metal windows, and a special compound should be obtained for the purpose. Windows should be glazed to their sight lines, that is, the edge of the metal frame.

Where panes of glass are extra large, as in picture windows or some types of door, it may be necessary to use plate glass rather than the standard thickness. The Building Inspector's Department will usually advise on this. Toilet and bathroom windows are glazed with obscure glass, and also the front and side doors. The most common types of glass for this purpose are the Arctic and Hammered. More attractive and only slightly more expensive are the Spotlyte and Cross Reeded types.

Specification clauses

The specification clauses for the Bricklayer and Concretor are the same as those in the previous chapter, with the addition of Clauses Nos. 55 and 60 under Bricklayer.

Under Joiner clauses 63 to 68 inclusive and the part of Clause 74 dealing with the trimmings for the trapdoor are applicable.

The various flashings are specified under Plumber, Clauses 84 to 87 inclusive, whilst down pipes and gutters are mentioned in Clauses 104 and 105.

Should Finlock guttering be used then the specification will have to be modified accordingly.

Under Tiler, all clauses, 107 to 114 inclusive, are applicable, and also all clauses under Glazier, being 130 to 132 inclusive.

Quantities of materials

The quantity of brickwork for this final stage will be similar to that in Chapter 5, with the addition of 1000 facings and 1000 commons if a gable-end is involved. The quantity of common bricks will be modified slightly if the internal walls do not continue on up and breeze-blocks and studding are used instead.

With this final quantity it is now possible to get an overall picture of the total quantity of bricks required. This brings the amount of facings to approximately 10,000, and the commons to 19,000. This figure will naturally vary one way or the other depending on the size of the building.

Quantities of sand and cement will be slightly more than those quoted in Chapter 5.

A small amount of $13\frac{1}{2}$ in D.P.C. will be required for any windows in a gable-end which have brickwork above them. One roll should be sufficient. First-floor windows do not usually need this treatment, as there is usually no brickwork above. Of course, all windows on the first floor require a vertical D.P.C. where the cavity is bridged. A further two rolls of $4\frac{1}{2}$ in. material should cover this.

Timber for the roof is a major item, and the easiest way of obtaining the requirements is to draw out the roof to a scale of $\frac{1}{2}$ in. to the foot. A section through the roof should be drawn first and then the plan view. It must be remembered that the rafters drawn in the plan view are not the true length, and this can only be obtained from the section. These drawings will also enable quantities to be obtained for wall-plate, ridge, purlins, soffit, fascia, struts and hangers.

For the roof covering itself, whether small tiles or the large concrete interlocking ones, it is fairly easy to estimate quantities. The manufacturers give data which states the number of tiles required for a square (10c sq. ft.) of roofing. In any case manufacturers will give an estimate if the plans are submitted to them. Data is also available for the quantity of tile battens required and the number of rolls of felt can be estimated from the roof area and the area of the individual rolls.

If Finlock guttering is used, a special form is supplied, and when all the details have been filled in, an estimate is supplied by the company.

Quantities for down pipes and ordinary gutters can easily be estimated from the drawing. It is not possible to standardise dimensions for the chimney-flashing, as these will vary with the size of the chimney and the pitch of the roof. An estimate must therefore be made of the developed sizes of the various pieces, and the total of these will give the overall size of sheet required. It should be remembered that there may be two weights of sheet lead involved, and the quantity should be split up accordingly.

Finally there is the glazing. The glass will mainly be 24 oz. sheet glass, except where a special obscure glass is required or in the case of an extra large window. In the latter case $\frac{3}{16}$ in. plate is usual. Glass is cut to size before delivery and is obtained by quoting window-frame references.

Costs

Brickwork costs will be very similar at this stage of the work to those quoted in the previous chapter, being in the region of £74·00 for facing bricks and £72·00 for common bricks. The cost for common bricks will be adjusted slightly if blocks are used for some of the first-floor walls.

Windows could again be similar, say £35·00, but this will obviously vary depending on the window arrangement and the type used. Glazing the windows would cost between £35·00 to £40·00, which includes both labour and materials.

Roof timbers are a major cost for this section of the work, and for the type of house which we are considering the cost will be around £50·00. The roof timber costs for a bungalow would probably be greater, but of course there are no first-floor joists, etc., to consider, so the overall costs hardly vary.

Roof covering is another large item, and should be in the region of £75·00 including felt and tile battens. Fixing costs should be about £20·00 for the average-size house with a straightforward pitched roof, and using concrete tiles.

A cost of £10·00 is an average figure for gutters and down pipes, but it should be borne in mind that if Finlock Guttering is used, some of these items will not be required, and it is possible to make a total saving on the house. It should also be realised that by using this type of guttering there is a continued saving in as much as there is no subsequent painting after the house is occupied.

The other costs are fairly straightforward and are listed in the average summary.

Quite a considerable saving can be made in various instances, and these can also be seen from the summary.

The cost of fixing roof timbers could be saved by doing the job oneself or possibly halved by assisting the carpenter.

The fixing of gutters and down pipes is very straightforward and this cost can easily be saved.

Where glazing is concerned, it is probably better to have the job done professionally, especially where large pieces of glass have to be put into first-floor windows.

Average summary

COSTS

	As built £ p	Using own labour £ p
5300 facing bricks @ £14·00 per 1000	74·00	74·00
7400 common bricks @ £9·70 per 1000	72·00	72·00
Cost of laying—£15·00 per 1000	190·00	—
Cost of laying—£12·00 per 1000	—	152·00
Cement for bricklayers—32 cwt.	17·60	17·60
Sand for bricklayers—6½ yd.	13·00	13·00
Door hood	25·00	25·00
Roof timbers	50·00	50·00
Labour for fixing	25·00	—
Tiles, felt and battens	75·00	75·00
Fixing tiles	20·00	—
Gutters and down pipes	10·00	10·00
Fixing gutters and down pipes	10·00	—
Metal windows	37·50	37·50
Glass and putty	25·00	25·00
Labour for fixing	10·00	—
Flashing (lead)	15·00	15·00
Flashing (fixing)	15·00	—
D.P.C.	1·80	1·80
Flue liners	20·00	20·00
Sundries	20·00	20·00
Total	£725·90	£607·90

Seven

Services—hot and cold water—electricity—gas

HOT AND COLD WATER

As well as the Building Regulations there is also a set of Water Bye-laws to be taken into consideration. Individual Water Boards base these on a set of model bye-laws issued by the Ministry of Housing and Local Government, and the main items which are applicable to the installation of hot and cold water systems are detailed below.

The supply pipe from the Water Board's main (known as the service pipe or rising main) should be laid at a minimum depth of 2 ft. 6 in. The rising main is not allowed to pass through any man-hole, cesspit or septic tank. Once this pipe comes beneath a building then the depth at which it is laid can be less than 2 ft. 6 in.

Every tank and water fitting should be positioned within the building so that there is no risk of damage by frost. If an outside tap is to be installed, then it must be suitably insulated and waterproofed.

All fittings inside the building should be placed in such a position that they are readily accessible for replacement or repair. This means that the practice of embedding pipes in the wall-plaster is not correct. If it is desired to hide any runs of water-pipe, they should be enclosed in a duct designed for the purpose.

The Water Board must be allowed to fit a stop-valve on the rising main. This is usually done just outside the boundary wall

9. First-floor joists in position

10. Wall-plate in position

11. View of roof timbers and stack

12. Internal view of roof timbers, showing rafters, ceiling joists, strut and purlin

13. Typical chimney-stack flashing detail (*Courtesy of Lead Development Association*)

of the property and the covered box giving access to the valve is installed flush with the pavement. The rising main should also have a stop-valve just after its point of entry into the building. A stop-valve should also be fitted to every draw-off pipe from each cold water storage cistern.

All cold water storage cisterns and flushing cisterns should be fitted with an efficient warning pipe and no other overflow. The diameter of this warning pipe should be greater than the internal diameter of the inlet pipe. In any case it should not be less than $\frac{3}{4}$ in.

One other important point in these bye-laws is that only a flush-pipe from an approved flushing apparatus is allowed to deliver water to a water closet. In other words it is not permissible to couple the main supply direct to a W.C.

The points mentioned here are the main ones in the bye-laws. Several other clauses are automatically complied with by the purchase of approved fittings which are manufactured to various British Standards.

In addition, it may be that the Water Board will require such fittings as taps, hot and cold water storage cisterns, etc., to be tested by them at their depot and to be approved and stamped to this effect by them. This point can be checked with the Authority and it is also a good idea to obtain from them a copy of their bye-laws.

Information and detail on plans

The plans should show the position of the various pieces of equipment connected to the hot and cold water system. That is, the ground-floor plan will show the kitchen sink, domestic boiler and any W.C. that there may be. The first-floor plan will show the positions of the W.C., the bath, lavatory basin and the hot water storage cylinder. Whether it is necessary to show any of these items on the section will depend on where the latter has been taken. The additional item which could show on this view is the cold water storage tank in the roof.

Constructional details

Although the services have been left until this chapter, water will have been required on the site right from the initial start of the building work. This means that there must be some form of temporary supply in order to obtain water for building purposes. On most normal sites it is the custom for the Water Board to run a service pipe to the boundary wall where a stop-tap is fitted. The continuation of the service pipe or rising main from there into the building is the responsibility of the building contractor. In this case, of course, the building contractor is the reader.

In order to give a temporary supply of water, the Water Board can complete their portion of the work and a length of service pipe can then be taken from the stop-tap to a stand-pipe.

The water main is usually under the footpath, and the Water Board may or may not allow the person requiring the service to expose their main. If the footpath has already been made up, then the main will have to be exposed by their own labour. In cases where only the roads and services have been laid on a site under development, the Authority may allow the contractor (or reader) to dig down and expose the main, thus making the labour charges slightly less.

In order to make a connection for the service pipe, the Water Board must drill and tap their main. This can usually be done without the water being turned off, and whilst the main is under pressure. A special drilling machine is used for this purpose which is securely held to the main whilst it is being tapped. It is so designed that it allows the drilling and tapping operations, and also the insertion of the ferrule to which the service pipe is connected, to be made under pressure.

The length of service pipe is then taken to a stop-tap at the boundary of the site. It is then usual to build a dry-brick duct which terminates in a cast-iron box at pavement level. It is essential that the position of this access box is checked very carefully to ensure that it does not foul the boundary wall when it is erected at a later date.

The service pipe is then taken from the Water Board's stop-tap to a tap on the site itself, and to a position where it is most convenient for building purposes. The final run of the rising main into the building can be done at a later and convenient stage. When this rising main is finally run it must be remembered that it must be at a depth of 2 ft. 6 in. below ground.

One point that may be worth considering is to lay the rising main in the foundation trench. This has the advantage that one always knows where the pipe is, and by using the foundation trench a certain amount of excavation is eliminated. If it is decided to do this, it is usual to place a sand-bed on top of the footings and between the wall of the building and the edge of the trench. The local Water Board should be consulted on this point and it is most likely that they will want to come and inspect this part of the installation.

It is not possible to do this, of course, if the foundation trench is not deep enough to give the required cover to the pipe.

Putting the rising main in the foundation trench is usually done when brickwork has reached D.P.C. level and just prior to the trenches being back-filled. Although this takes the water supply right into the building, it does mean that some care must be taken in drawing off building water. As the tap is inside the building, water must not be allowed to run onto the floor concrete or brickwork. If water is being drawn off *via* a tap and a length of hose the latter should be securely fastened. If there is any danger of the tap being interfered with whilst there is no one on the site the water should be turned off every day at the stop-tap.

The portion of the rising main which runs underground from the stop-tap to inside the building is laid in soft copper pipe. The usual diameter for domestic purposes is $\frac{1}{2}$ in.

We now come to the hot and cold water system itself. This consists of a kitchen sink, bath, lavatory basin, W.C., cold water storage tank, hot water storage cylinder, a boiler, all the interconnecting pipework and taps. To understand the working of the hot water part of the system it must be realised that water, when it becomes heated, is less dense than cold water, and therefore it rises. Fuel is supplied to the boiler, and the heat generated is then passed by conduction to the water in the boiler which is at the very bottom of the system. The boiler itself has two pipes, one at the top and one at the bottom, which are led away to a copper hot water storage tank, which is usually, but not always, situated in the bathroom.

As the water in the boiler is heated it becomes less dense, and rises up the primary flow-pipe (which is the one at the top of the boiler), and is led into a point near the top of the hot water storage tank. As this water rises, cold, and therefore denser, water comes out from the bottom of the storage tank, down the primary return, and into the bottom of the boiler ready for heating. Thus it will be seen that there is a continuous movement of water due to convection, and the complete volume of water will, in this way, become heated.

Having obtained hot water, there must be some method of making good water which is drawn off. This is done by the cold water storage tank in the roof space. The rising main is taken from the stop-tap at its point of entry into the house, and is run straight up to the cold water storage tank. The rising main feeds this tank *via* a ball valve. There is then a feed from the bottom of the cold water storage tank, direct to the bottom of the hot water storage tank. This run of pipe should be fitted with a stop-tap.

The system must be vented and this is done by means of a vent-pipe which runs from the top of the hot water storage cylinder up to the cold water storage tank. It does not, however, connect directly to this, but is taken up, bent over, and left open ended above the water surface. The expansion pipe serves a dual purpose of allowing air to escape when the system is filled, and also allows the release of any pressure whilst the system is in operation.

Although it is not always realised, the expansion pipe of a hot-water system does contain water, and this is due to the fact that the system is under pressure. The pressure, of course, is due to the head of the water derived from the height and position of the cold water storage tank. The level of the water in the expansion pipe will be higher than the level of the water in the cold water storage tank. This is due to the fact that the head of water is given by heavier, denser water than that in the cylinder and the expansion pipe. The point of taking the expansion pipe up high enough before bending over to the storage tank then becomes apparent. Due to the fact that the expansion pipe contains water, it is from this source that the hot water supply itself is drawn. A tee is placed in the expansion pipe just above the top of the hot-water cylinder, and the hot-water feed for the bath, lavatory basin and kitchen sink is taken from this point. Cold water for the bath and lavatory basin is taken from a junction made in the cold-water feed from the cold water storage tank. The cold water supply to the kitchen sink is taken from the rising main on its way up to the cold water storage tank. This item is a requirement to ensure that there is an uncontaminated supply of water for drinking purposes.

The foregoing describes a system known as the 'direct' system. If it is intended to instal a number of radiators, it is advisable to use what is known as the 'indirect' system. With the indirect system the hot water storage tank has another cylinder inside it, and the primary flow and return are connected to it. Pipework for the radiators is also on this primary flow and return. This means that the water in this part of the system is, apart from the cold-water feed and vent-pipe which is still necessary, flowing in a closed circuit. The water in the inner cylinder, or calorifier as it is known, in turn heats the water in the storage tank, and it is this water which is used for the hot supply.

By using this method the same water is continuously circulating through the boiler and the radiators, and consequently this minimises the deposits of scale. A certain amount of make-up water will be required and this can either come from the main cold water storage tank or the Local Authorities may require this feed to come from a separate tank.

Dealing with the individual components themselves, the first item to be considered is the boiler. The normal type of boiler used for a domestic hot water system is a free standing unit of about 20,000–25,000 B.T.U.s per hour capacity. The boiler itself should be bower-barffed in order to make it rustless. It should also be fitted with a drain-cock. There will be connections for the primary flow and return, probably for $1\frac{1}{4}$ in. pipe. Care should be taken to ensure that the connections are obtained on the correct side of the boiler. As mentioned in Chapter 9 the boiler can stand on a quarry-tiled plinth. When the boiler is in position the vitrious enamel flue-pipe can be fitted. The joint at the top of the boiler is made with fireclay and the plaster, etc., must be made good where the pipe enters the flue either *via* the wall, lintel or register plate. The soot door, necessary with a free-standing boiler so that the chimney may be swept without any dismantling, will have already been built into the brickwork. In addition a soot door is sometimes incorporated in the flue-pipe itself.

The pipework from the boiler is then taken up to the appropriate connections on the hot water storage tank. In the case of a house this tank usually stands directly on the floor-boards in the airing cupboard, but in a bungalow it is usual to mount it on a wooden support.

If it is desired to use an immersion heater as an alternative method of heating the water, the tank should be obtained with the appropriate connection. To complete this part of the installation the tank should be lagged.

The cold water storage tank is positioned in the roof space and should be against a warm chimney-breast. It is usual to have a galvanised tank which is in the region of a 50 gal. capacity. This means that there will be a weight of 500 lb. in the roof space, and it is essential to see that this load is well supported. The tank should be mounted, supported on the appropriate timbers, so that it comes over or near a load-bearing wall. A point that should be checked is to ensure that the size of tank chosen is not too big to go through the trap opening.

While the traditional material for storage tanks is galvanised iron there are also available storage tanks manufactured from fibre-glass and PVC. These materials have the advantage that they do not deteriorate in use, a factor which is well worth bearing in mind before a final decision is taken on this part of the installation.

The rising main is connected to the tank *via* a high pressure ball-valve. An overflow pipe must be taken from the tank, through an outside wall to discharge in a conspicuous position. The cold-water feed is taken from the bottom of the cold tank, down to the bottom of the hot-water tank, usually in ¾ in. pipework. This

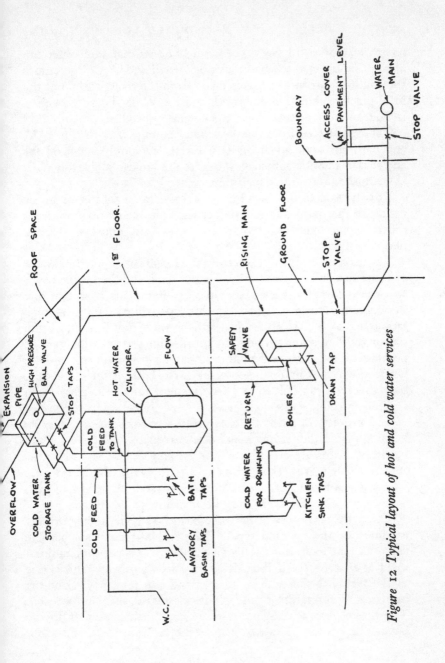

Figure 12 Typical layout of hot and cold water services

length of pipe will have a Tee connection so that cold water can
also be supplied to the bath, lavatory basin and W.C. Sometimes
the cold feed for these items is taken from the rising main, but the
local practice should be followed, and the Water Board consulted
if there are any doubts about the correct procedure.

Taking the feed from the cold water storage tank for the W.C.
and bath has some advantages. If the water supply is cut off for
any reason there is sufficient water in the tank to allow the W.C.
to be flushed for quite a long time.

This type of feed is also better when using a shower or mixer
fitting in the bath, due to the fact that the mains pressure is so
much higher than the pressure in the system, and even mixing
becomes difficult.

The expansion pipe is taken from the top of the copper hot-water
cylinder to a point above the cold-water tank, and then turned over
and down towards the surface of the water. This is so that any
discharge that there may be from this pipe will go into the storage
tank itself. A Tee is placed in the expansion pipe just above the
top of the hot-water cylinder to supply hot water to the various
points required. The vent pipe itself can be $\frac{3}{4}$ in. diameter and the
hot feed will be $\frac{3}{4}$ in. diameter as far as the bath. After serving the
bath this feed can reduce to $\frac{1}{2}$ in. diameter for the lavatory basin
and kitchen sink.

The majority of baths which are fitted in houses are 5 ft. 6 in.
long. A slightly larger one may be preferred and a 6 ft. one could
be fitted. Baths are on adjustable legs, and when the bath is installed, the legs are set to give a slight fall towards the bath waste.
The waste is fitted with a 3 in. deep-seal trap and the size of pipe
is usually $1\frac{1}{2}$ in. diameter. A $1\frac{1}{4}$ in. overflow is also fitted.

There are two types of lavatory basin available, either bracket
mounted or the pedestal type. The pedestal-type basin has the
more elegant look, but its installation may require some adjustment at a later stage if it is fitted on a wooden floor. This is due
to the fact that the timbers shrink, and the tendency is for the
basin to be supported by its pipework rather than its pedestal.
The waste-pipe attachment is a trapped fitting, similar to the
bath. There is no separate overflow pipe, however, as this is taken

care of in the design of the basin. Should the basin overflow, this water is by-passed *via* a built-in channel back to the main waste.

Readers who are looking for a degree of elegance in their bathroom may feel inclined to make a particular feature of this part of the house. If this is the case an attractive wash basin unit with a laminated plastic top could be considered or, while not so elaborate but still in the modern idiom, a wall-mounted basin with concealed attachments would be an attractive alternative.

The W.C. is installed by screwing the pan to the floor and connecting the outlet, which is in the form of a trap, to the branch of the soil and vent pipe as described in Chapter 8. The flushing cistern is then screwed to the wall and the cold-water feed is connected to the ball valve inlet. The ball valve from the W.C. should be either high or low pressure depending on whether the mains or storage tank is used as a feed. A further $\frac{3}{4}$ in. diameter overflow pipe must be fitted. The flush pipe itself is connected from the bottom of the cistern to the inlet of the pan.

The remaining item to be considered in the plumbing system is the kitchen sink unit. There are several varieties available and which is chosen is largely a matter of personal choice. Some readers may prefer to install a complete unit, while others may prefer to install a sink and integral draining board and then build up their own unit at a later date, matching it to their overall kitchen scheme.

The height at which the sink is installed is important, as it is a working area of the kitchen. The top of the sink should be approximately 3 ft. above the floor. This height should allow washing, etc., to be done without discomfort. Together with the height of the sink, the position of the taps should be given some consideration. They should be high enough above the sink to allow such things as buckets to be used.

The cold water tap at the kitchen sink has a direct connection to the rising main (to ensure a supply of pure water for drinking and domestic purposes) and the hot tap is connected to the hot feed from the hot water storage tank.

The waste pipe from the kitchen sink is allowed to discharge into a gulley, providing that the discharge takes place above the level of the water seal in the gulley, but below the level of any grating covering it.

Regarding the installation of the plumbing system, it is a job which could be undertaken by the reader, providing the necessary care is taken. The majority of modern domestic hot- and cold-water systems are executed in copper pipe. The piping used for the service pipe between the main and the house will be soft copper, whilst the rest of the piping inside the building will be in half-hard copper. There are two main types of fittings which are used with copper tube. These are compression joints or capillary soldered joints.

For the person who has a very limited amount of equipment, compression joints, which only require the ends of the tube to be cut square, are probably the easiest to use. The joint is made by fitting a loose sleeve over the end of the pipe and then making the joint by tightening the coupling nuts. There is a further type of compression joint which requires the tube to be belled at the end before the joint can be made. Compression joints do, however, have the disadvantage that they have a rather heavy and clumsy look.

Capillary soldered joints have a more pleasing appearance, and this is because they are little larger in diameter than the pipe over which they fit. These fittings rely on the capillary action of the molten solder which is drawn into the annular space between the inside diameter or the fitting and the outside diameter of the pipe. These joints have to be made with a blowlamp and the solder can either be applied at the end of the fitting or through a touch hole. There are also some fittings on the market which contain a supply of solder and this is caused to flow and make the joint when heat is applied.

It is possible to completely fabricate a water system without any of the pipework being bent. If this is done the swept or gradual radius bends should be used, rather than the sharp elbow bends. The bending of copper pipe is something which should not be undertaken by the reader unless he has access to the proper equipment. Bending is done by either using a small bending machine containing the necessary formers, or by inserting springs in the pipe before bending. Both methods ensure that the pipe does not close up when it is manipulated.

Whilst on the subject of pipework it is necessary to mention a material which is being used increasingly for this type of work. This is polythene, and both pipes and fittings are available in it. Polythene piping is at present only suitable for cold water installations, although a development may become available for domestic hot water systems also. From the reader's point of view polythene piping has the advantage that it can be bent cold, although sharper radii can be obtained with hot bends. On the other hand, the material is not so elegant in appearance as copper pipe and it also needs careful support to prevent sagging.

Before deciding to use polythene piping for cold water plumbing, the Local Authority should be consulted to ensure that this material is acceptable to them.

Before leaving the subject of domestic hot water it should be pointed out that the system described here has referred to domestic hot water heating in its simplest form. Today the accent is on central heating in one form or another, and the reader should consider the pros and cons of this subject very carefully before coming to a final decision, bearing in mind the fact that an increasing number of new houses are now being built complete with central heating. This is, of course, a big selling point, and in future years houses without this feature will be at a considerable disadvantage.

When deciding on the domestic hot water installation the reader is faced with three possibilities: he can install a free-standing boiler or back-boiler unit which will provide his basic domestic needs, space heating being provided by other means; he can install a free-standing boiler or a back boiler unit which will also supply a number of radiators, again making up the extra heat needed by other forms of space heating; he can install a boiler large enough to give both domestic hot water plus an increased heating load.

Of course, if some other form of space heating is required, then the problem usually becomes one of providing domestic hot water only, or with possibly the addition of a limited amount of radiator heating for background warmth. And, when the latter types of installation are being planned, it is usual to use an indirect system where the hot water cylinder has a tank or coil inside it connected to the primary flow and return of the boiler. This system ensures that the water is never drawn off or replaced and this cuts scale deposits to a minimum.

Where a number of radiators are employed for space heating it is usual to have the system pump assisted. This enables small bore pipes to be used, which are relatively simple to install and unobtrusive in appearance.

It should be realised that oil fired boilers can be used for combined hot water and space heating installations. The cost of such a boiler, together with its oil storage tank and various controls, is more expensive than a solid fuel boiler. However, the reader may prefer to install this type of boiler and effect a saving elsewhere in the building of the house.

Should an oil-fired boiler be contemplated, the reader will find that the various oil companies concerned make available a large amount of technical information and advice.

ELECTRICITY

Before going any further it must be pointed out that unless the person building the house is a qualified electrical engineer, the installation of this equipment should not be undertaken. Electricity can be lethal, and it is not worth running either the risk of an accident or fire by trying to economise with this part of the work.

There are no bye-laws, as such, covering the installation of the electrical supply and equipment. The requirements for this installation, however, are laid down by the Institution of Electrical Engineers, and are published in a book entitled *Regulations for the Electrical Equipment of Buildings*.

The main points in these regulations, as far as the householder is concerned, deal with cable sizes, fuse ratings, the switching of lighting fittings and the installation of switches and electrical equipment in bathrooms.

Information and detail on plans

It is not usual to indicate any details of the electrical requirements
on the plans submitted to the Local Authorities, other than to
mark the position of a cooker and wall fire. When the time comes
to obtain quotations for this work the usual practice is to give a
marked-up print to the electrical contractor concerned. On this
print will be shown the lighting and switch positions, sockets,
outside lights and the position of any fixed appliances.

Installation details

In recent years there has been considerable development in the
design and installation of electrical equipment. There is no
longer the assortment of plugs that there used to be, and the
wiring itself is less complicated and consequently far cheaper.

As it will be appreciated, a large proportion of the work involved
in wiring up a house must be done before the plastering is started,
and the plasterboard for the ceilings is put in position. In some
cases part of the work may even be undertaken before the floor-
boards are laid. After the cables are in position great care should
be taken to ensure that they do not suffer damage through subsequent nailing of floors and ceilings or through any other cause.

Before the plastering is started, the brickwork should be chased
out where necessary to accommodate the boxes of socket outlets
and cooker units. Switches themselves are now obtainable to suit
the plaster depth, and no chasing out is necessary for these items.
The cables for the various circuits are embedded in the plaster
and protected by a light galvanised capping, which is attached to
the wall at intervals and forms a complete cover over the cable.
By using flat cable and this capping, the need for conduit and its
attendant chases is eliminated.

Domestic electricity supply is now standardised at 240 volts
A.C. The use of alternating current instead of D.C. or direct
current has played a large part in the improvements that have
taken place in the design of equipment.

The Electricity Board will be responsible for bringing the supply
into the house. The cable will enter the building through a duct as

described in Chapter 4. The Electricity Board should be consulted as to the position and depth at which their cable will enter the building. The supply or service cable terminates in a service fuse and neutral link and then the meter. This equipment is the property of the Electricity Board. The supply from the meter then goes to the main isolating switch and distribution board. From this distribution board, or consumer's unit as it is known, are taken the various sub-circuits. The consumer's unit is obtainable with a varying number of ways, but a six-way unit should be suitable for the average sized domestic building.

These sub-circuits would probably consist of two 5 amp. circuits for lighting, one 30 amp. cooker circuit, one 30 amp. ring circuit and one 15 amp. water-heater circuit. This leaves one spare way available for any circuit which may be added in future.

Another method is to run the power circuit from a 30 amp. switch, and the lighting circuits from a separate switch splitter. This is a switched unit which enables two lighting circuits to be taken from it. With this method there are no allowances for future circuits.

The cables used on these circuits will vary in size, according to the load which they are called upon to carry. The cable used will either be Tough Rubber Sheathed (known as T.R.S.) or Plastic Covered (known as PVC). On the ring, cooker and water heater circuits the cable will have an internal earth wire, and an earth continuity conductor is also recommended for lighting circuits so that an earthing point is always available when metal fittings are used.

In a house two lighting circuits are used, one for the ground floor and one for the first floor. The cable for this will probably be 3/·029. This means that each conductor has three strands of wire, each ·029 in. dia. Three-terminal ceiling roses are used (with the exception of the last one on the circuit) and the lighting points are wired in parallel. If any fluorescent fittings are used they should be earthed. In kitchens and bathrooms it is usual to instal spherical fittings which completely surround the bulb and the bowl screws directly to a base fitted on the ceiling. In bathrooms the switch should be either a cord-operated ceiling switch

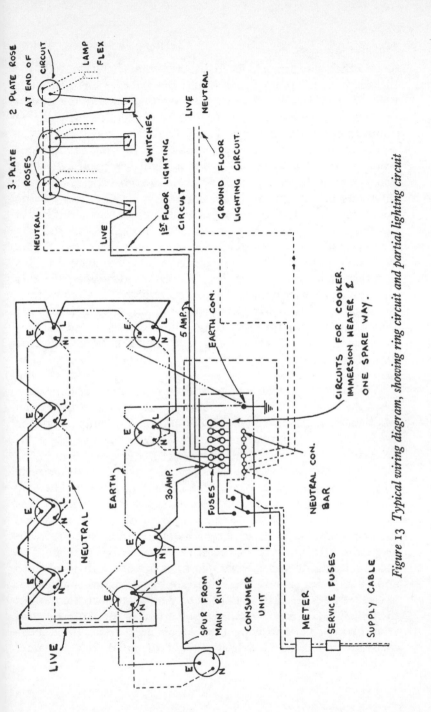

Figure 13 *Typical wiring diagram, showing ring circuit and partial lighting circuit*

LAMP FLEX

2 PLATE ROSE AT END OF CIRCUIT

3-PLATE ROSES

SWITCHES

NEUTRAL

LIVE

1ST FLOOR LIGHTING CIRCUIT

GROUND FLOOR LIGHTING CIRCUIT.

LIVE

NEUTRAL

E N

E N

E N

E N

E N

E N

E N

E N

5 AMP.

EARTH CON.

EARTH

NEUTRAL

LIVE

30 AMP.

FUSES

3OAMP.

NEUTRAL CON. BAR

CIRCUITS FOR COOKER, IMMERSION HEATER & ONE SPARE WAY.

SPUR FROM MAIN RING

CONSUMER UNIT

METER

SERVICE FUSES

SUPPLY CABLE

or else the switch should be positioned immediately outside the bathroom door. Whilst considering the lighting circuits, some thought should be given to external lights. Probably the neatest unit for an external light is the bulkhead fitting, which can be mounted directly to the wall. This fitting is weatherproof.

The other main circuit in a house will be the ring circuit. This will be installed in 7/·029 cable with a bonded earth. If a dwelling does not exceed 1000 sq. ft. in area, and provided it is known that the rating of any appliance will not exceed 13 amp., then a ring circuit may be installed which has an unlimited number of socket outlets. In addition it is permissible to take spurs off the ring main provided the spur does not serve more than two socket outlets, and provided that the spurred outlets are not more than half the total number of outlets. As far as practicable, the ring circuit should, as its name implies, go right round the building. Spurs are used to reach parts of the building to which it is not convenient to run the main ring. They are also used in conjunction with a spur fuse-box for such items as built-in electric fires. The ring itself feeds both ways and there will be a 30 amp. fuse in the consumer unit.

While one ring circuit is suitable for a small house or bungalow, when the floor area exceeds 1,000 sq. ft. it is necessary to have an additional circuit. If this is the case the two circuits should be evenly balanced, with approximately the same number of outlets on each.

Socket outlets are now shuttered, the shutter being pushed aside when the plug is inserted. For use in conjunction with the ring circuit the 13 amp. fused plug has been developed. These plugs have flat pins to suit the socket and each plug carries its own cartridge fuse. When these plugs are obtained they are fitted with a 13 amp. fuse, but it is important that every appliance should be fused correctly. Separate cartridge fuses are available in 2, 5, 10 and 13 amp. ratings, and it is important to see that each appliance is correctly fused. If this is not done and a fault occurs, the flex to the appliance may be called upon to carry a far greater load than it is capable of withstanding. As a general guide, fused plugs to lamps and small appliances should be fitted with a 2 amp.

fuse, irons and medium appliances with a 5 amp. fuse, while 10 and 13 amp. fuses should be used with electric fires and large appliances. In addition it should also be checked that the size of the flex to each appliance is adequate in size.

When planning the ring circuit a great deal of thought should be given to the number of socket outlets required. The general tendency is to allow far too few. Outlets should be allowed for in the hall and landing for vacuum cleaners. In addition, in view of the various electrical appliances now available, it would not be out of order to have a minimum of six in a living-dining room, and at least four in a double bedroom.

If an electric cooker is to be used, it will be on its own circuit and the cable used will be 7/·044. A cooker is controlled with a cooker unit and this usually has a socket outlet, which can be used for an electric kettle.

The foregoing applies to normal domestic power and lighting circuits. If electricity is chosen as a means of central heating, utilising off-peak power, then the storage units installed will be on individual sub-circuits connected to a separate consumer's unit and meter.

Electricity may also be used for Electricaire units, in which a storage heater supplies warm air through a ducting system, and for under-floor heating. Here again a separate meter is used, but only one sub-circuit is needed.

When all the wiring is complete the Electricity Board must be notified, and they will come and check the installation before making the final connection to the mains.

GAS

When gas appliances are being used it is usually necessary to fit service pipes before plastering is commenced. Gas is another service which can be lethal, and it is advisable to leave the installation to the Gas Board.

A service pipe is brought from the Board's main, into the house, and terminates at a meter. Pipes are then run from this to the

E

various points where gas is required. A gas cooker is usually the minimum requirement but, like the other services mentioned, gas is also popular for use with both 'wet' and 'dry' central heating installations. If the maximum use is made of this service, then the Regional Gas Board will advise on their requirements for the particular appliances which are to be used.

When the pipework is installed it will be laid with a fall back to a point from which condensed moisture in the gas can be drained. The entry of the service pipe into the building should not be through the same duct that the electric cable is in.

Specification clauses

The clauses in the sample specification which cover the services are 88 to 99 inclusive and 106 under Plumber, and clauses 115 to 120 under Electrician.

Quantities of materials

Once again, the quantities of materials required to install the service is a question of direct measurement on site. In addition, such items as baths, sinks, etc., are obvious, and it is not difficult to compile such a list to suit personal requirements.

In obtaining equipment and material for this stage of the work there is also the advantage that most of it can be viewed at the local builder's merchant, so that the exact requirements can be suited.

Costs

The costs for the services are fairly straightforward, but a check should be kept to see that one is not over spending on such things as baths, elaborate fittings, cookers, etc.

For the water service, the prices quoted are for an average system.

Coloured bathroom suites can be an expensive item, especially if coupled with an elaborate design.

If a separate toilet is installed it may be worth considering having a coloured bath and basin, whilst the W.C., on its own, can be white.

Mixer taps and shower fittings are also items which can make the cost mount up, but if considerable savings have been made elsewhere the fitting of these may come within the budgeted allowance.

On the labour side it is possible to save the complete cost if it is decided to undertake the work.

With the electrical and gas services there is little that can be done to lessen the costs. Both services should be left to those qualified to deal with them, and any attempt at making a saving on this could be false economy.

Average summary

	COSTS	
	As built	Using own labour
	£ p	£ p
Water for building	8·00	8·00
Connection to main	20·00	20·00
Domestic boiler	35·00	35·00
Stainless steel sink top and bowl	12·00	12·00
Sink trap	1·00	1·00
6 ft. coloured bath	40·00	40·00
Pedestal lavatory basin—coloured	14·00	14·00
W.C. low-level suite—white	12·00	12·00
Bath and L.B. waste, and bath overflow	2·50	2·50
Hot water storage tank—copper	16·30	16·30
Galvanized cold water tank	6·00	6·00
One coil of soft copper, $\frac{1}{2}$ in. dia.	11·00	11·00
$\frac{1}{2}$ in. dia. copper tube for internal work	8·50	8·50
$\frac{3}{4}$ in. dia. copper tube	7·00	7·00
$1\frac{1}{4}$ in. dia. copper tube	9·00	9·00
Fittings, clips, etc.	10·00	10·00

Pillar cocks for lavatory basin, H. and C.	2·75	2·75
Pillar cocks for bath, H. and C.	3·75	3·75
Bib cocks for kitchen, H. and C.	1·35	1·35
Labour for plumbing	25·00	—
Electric cable	26·00	26·00
Socket outlets, 20	11·20	11·20
10 fused plugs	2·00	2·00
12 ceiling roses, lamp-holders, etc.	4·25	4·25
Switches	3·50	3·50
Consumer unit	6·00	6·00
Galvanized capping	4·25	4·25
Labour	20·00	20·00
Gas cooker	50·00	50·00
Total	**£372·35**	**£347·35**

Eight

Drainage—manholes—soak-aways—septic tanks—testing

Building Regulation requirements

The main Building Regulation requirements for the drains are that they shall be of adequate strength, laid to the correct falls and provided with watertight joints. They must also be of adequate size and in any case drains connected to soil appliances (which includes a W.C.) must not be less than 4 in. diameter. And, after laying and the trench back-filled, the drains must be able to withstand a test to ensure that they are watertight.

Where the run of the drain takes it near a load-bearing wall, or passes through a wall, precautions must be taken to ensure that the stability of the building is not affected and that the drains will always remain watertight.

Manholes are required at various points and in particular where any sewer changes direction.

Manholes should allow access for rodding and be constructed to exclude subsoil water. In private dwellings this is met with by building the manholes in 9 in. brickwork. In addition the bottoms of the manholes should be finished with channels properly benched and the top should be protected by a cast-iron cover.

A further requirement is that any inlets to a drain shall be protected by a trap or water-seal, with the exception of the soil pipe. Here it should be noted that the definition of soil pipe is a pipe (other than a drain) which conveys either soil water alone or soil water with either waste water or rainwater, or both.

In the case of a normal house or bungalow, where it is usual to incorporate a 'one-stack' drainage system, the soil pipe will extend upwards and become a ventilating pipe also—usually referred to as the soil and vent pipe. The individual waste pipes from baths, washbasins and sinks then flow into this.

In order to protect the drainage system from freezing up, and also to generally improve the external appearance of the building, the Building Regulations require that all soil and waste pipes are situated within the building. However, an exception is made for the waste discharge from a ground floor sink which may, if required, discharge into a trapped gulley, providing that this is effected above the level of the water in the trap but below the level of the grating.

A further requirement is that the ventilation pipe itself must be carried up to such a height and position where escaping foul air will not be a nuisance or a health hazard. It must also be topped with a wire cage which must admit the passage of air but prevent birds nesting. And, after jointing, the soil and vent pipe must be capable of withstanding a smoke or air test.

There are also regulations concerning the construction of the water-closets themselves. The pan itself must be properly shaped and have a non-absorbent surface which can be properly cleaned and provided with an effective flushing apparatus. No part of the pan shall be connected to other than the soil pipe and the flushing pipe.

In areas where cesspools and septic tanks are used, these must be of adequate construction and resist the passage of liquid either from inside or outside.

If the cesspool is not of the settlement type it must be properly covered, adequately ventilated and have means of access for emptying its contents.

If the cesspool is of the settlement type, it must be adequately ventilated and covered, and its discharge must not cause any form of pollution.

It should be noted that the Building Regulations state that a cesspool must have a capacity of not less than 4,000 gallons, while a septic tank must have a minimum capacity of 600 gallons. In view of the size requirements for cesspools it is obviously a cheaper proposition to construct a septic tank.

Information on plans

Drainage information as shown on the plans is fairly straight-forward. The positions of the gullies and W.C.s are shown on the plan view and the flow of the drains indicated by a single line from them to the main sewer or septic tank. This information is also included (with exception of showing the gullies) on the site plan, as it is not possible to show the complete run of the drains on the $\frac{1}{8}$ in. to a foot plan if the sewers are any distance from the house.

In the elevations the down pipes for the discharge of water from the gutters will be shown, and also the vent pipe where it goes through the roof.

Manholes are also shown on the plan view, and these will usually be where the drains change direction and immediately before entry to the main sewer.

Gullies should be noted on the drawings and also sizes of pipes and the material for the soil and vent pipe. The cage on the top should also be indicated and noted.

Constructional details

Drainage details vary in their complexity, depending on the distance to go to reach the sewers and whether there is combined or separate drainage.

Combined drainage means that there is only one sewer to accept both foul and storm water, whilst with separate drainage there is one sewer for each.

In country districts where no sewers are available it is usual to take the wastes into a cesspit or septic tank, and the rainwater into soak-aways.

What are the possibilities of the reader doing this work himself? Of course, the person who is tackling a whole house will do this stage of work, but the reader who is having the shell of the building constructed and therefore aiming to save as much as possible, should consider this item very carefully.

The work falls into two sections, excavation, and laying and jointing the pipes. Excavation is straightforward, providing care is taken with falls and levels. The laying and jointing is also straightforward, but it has to be done very carefully as the drains have to be tested by the Local Authority. Any joints which fail the test have to be cut out and re-made, and once again tested to the satisfaction of an inspector.

Even if it is decided to employ a professional drainlayer, it should be remembered that they are not immune from having several joints fail. On the other hand, the writer knows of one quite complicated drainage system which was undertaken by a person building his own house (a clerk), which had only a slight weep at one joint when the system was tested by the Building Inspector.

With care this work could certainly be undertaken oneself, and the subsequent saving can be quite considerable.

For the purpose of this chapter it will be assumed that a separate drainage system is to be installed, and that the two sewers are adjacent to one another.

Domestic drainage is constructed in 4 in. dia. salt-glazed pipes, the 4 in. being the internal diameter. The pipes are 2 ft. long and each one has a socket at one end, into which the adjacent pipe will fit. Both the spigot and socket of a pipe are grooved and unglazed to facilitate the making of the joint. And, assuming that the traditional pipes are being used, the other fittings such as gullies, channels and bends will also be salt-glaze ware.

In the earlier requirements it was mentioned that the drainage shall be laid to an adequate fall. This fall is actually worked out as being 1 in. in 10 diameters. Most house drainage is 4 in. diameter pipework, so the required fall to ensure that the drains are self-cleansing is 1 in. in 10 × 4 in., or 1 in 40.

It is usual to put a manhole immediately before the entry to the main sewers, and this manhole makes a convenient point from which to start the drainage. Domestic manholes are usually 18 in. × 24 in. internal dimensions, which means that with 9 in. walls the external dimensions will be 36 in. × 42 in. The excavation will have to be made slightly larger than this to allow the bricks to be laid.

If the system needs a double unit with the storm and foul man-holes side by side, then the size of the base concrete must be modified accordingly.

A 6 in. bed of concrete is put in the bottom of the hole, and this serves a dual purpose as the bottom of the manhole and the founda-tion for the 9 in. walls.

The outlets at either end of the manhole are the standard drain-pipe, and these, and any side entries, are built in as the brickwork proceeds.

Between these outlets, on the base concrete, an open channel is placed, and a slipper if a side entry is involved. These open channels have to be benched, which is the sloping surface to the sides of the open channel.

The benching should rise vertically from the sides of the open channels until it reaches a height equal to the top of the outlet pipes. It should then slope upwards to the sides of the manhole. This slope should be 1 in 6.

The majority of this benching is in ordinary concrete, but should be finished off with cement-sand mortar. The mix should be 1 : 1 and should be given a steel-trowel finish.

The brickwork for the manholes should be built up just below ground or path level, and the manhole cover, when bedded on top of the brickwork with cement mortar, should represent the required finished level.

Starting from the lowest manhole point the drain excavation can then proceed towards the house. The trench should be just wide enough to allow the pipes to be jointed (approx. 1 ft. 6 in. to 2 ft.) and should have the correct fall of 1 in 40.

The run of the trench should be pegged and marked with string. The correct fall can be arrived at with pegs, using a level and a straight-edge.

Pegs are knocked in the bottom of the trench at measured dis-tances from one another, and their tops levelled with the straight-edge and level.

Distances are then measured down from the tops of the pegs

and the bottom of the trench adjusted accordingly. If the required fall is 1 in 40 and the pegs are 10 ft. apart, then the pegs are marked in steps of 3 in. That is 0 in. (the top of the first peg), 3 in., 6 in., 9 in., etc.

When all trenches and manholes are satisfactorily excavated to the correct levels, the laying of the pipes can begin. Pipes should be laid to a line, care being taken that the correct fall is preserved and that the run is straight.

Drain-pipes can either be laid directly on the ground, or placed on bricks. If the former method is adopted the barrel of the pipe should be on firm soil, and a portion should be scooped away to accommodate each socket and sufficiently large enough to allow for jointing.

If the drain-run is placed up on bricks, these should be placed immediately behind each collar. This method has the advantage that the pipes are kept clear of the trench bottom, and this facilitates jointing in wet weather if water is lying in the trenches.

Yet a further method is to bed the pipes on dumplings of concrete and the pipes are tamped into them and lined up in the normal manner. If this method is used, the pipes should again be supported behind the collars.

Salt-glazed pipes are jointed with a gasket of tarred hemp and a 1 : 1 cement mortar mix. This gasket prevents mortar getting into the pipes and also tightens and centralises them. The gasket is caulked home and the remainder of the joint completed with cement mortar and finished off with a 45 degree fillet on the outside.

When each joint is made, a half-round scraper should be passed through the pipe to ensure that there is no mortar left inside to give a rough ledge with its subsequent troubles. The joints should be covered with damp sacking until they are set.

From the last manhole the pipe runs should continue back towards the house, passing through any other manholes that may be necessary.

The storm-water runs will terminate in 'S' trap gullies. These should be mounted on brick or concrete piers and at the correct level in relation to any paths, etc. The connection from the gully to the main drain run is then made with the various slow and knuckle bends available.

The soil run will terminate at the soil and ventilating pipe, to which it should be joined by a large radius bend.

It is a Building Regulation that the test on the drains by the Local Authority has to take place after the drain trench has been backfilled, although they do inspect the work to see that it meets the general requirements before this test takes place. Consequently, when traditional pipes and joints are being used, it would be a common sense precaution to give the drains a preliminary water test in order to avoid the trouble and expense of opening up the excavation at a later date if leaks are found when the Local Authority come to make their own test.

For this test the system is filled with water (using a hose) and then examined for any leaks.

The pipes are made to retain water by plugging their ends at the manholes with a special expanding stopper. They are then filled at the gully, the hose-pipe being inserted round the neck of the trap. Care must be taken to see that the pipes are completely filled with water. Any air trapped in the system may disperse slowly and cause the water level in the gully to drop and give the impression of a leak. The plugs are left in for about twenty minutes and the joints examined for any wccps, and the gullies for any drop in water level. This detailed examination cannot, of course, be given when the trench has been backfilled—a further advantage to be gained from a preliminary test.

When sections of drain that have no gullies have to be tested (i.e. between manholes) the lower end is plugged in the normal way, and a special expanding plug is put in the higher end. This plug has a funnel attachment on it, and this takes the place of a gully to enable the pipes to be filled and retain the water.

Back-filling is an operation that has to be done with extreme care, and the exact method employed for casing the pipes will depend to a large extent on local conditions. In some cases it may be in order to haunch the pipes with concrete, while in other instances the Building Inspector may suggest that the pipes should be surrounded by granular material (such as gravel) before the normal back-filling material is returned to the trench. This latter material should be well distributed in layers and should not contain hard lumps or brickbats.

When this back-filling is completed the Local Authority is contacted to arrange a water test (already described) and a smoke test for the soil and vent pipe. This latter test is done by plugging the top of the pipe and also inserting a plug at the nearest manhole. The latter plug is connected by a flexible hose to a smoke machine, which is a device for burning oily waste, with bellows to drive the smoke into the pipe. To enable this test to be carried out all the connections must be made to the soil pipe and the traps filled with water.

As well as being responsible for these tests, the Local Authority is also responsible for the connections to the main sewers.

Notification must be given in good time and it is a good idea if the connection can be made before the test on the drains takes place. This will ensure that there is a getaway for the water lying in the pipes when it is released at the end of the test.

The actual connection is made by enlarging a small hole in the sewers and then concreting a saddle over the top of the final hole. The saddle is then connected to the last manhole with ordinary pipes and bends in the normal manner.

Before leaving the subject of drainage systems, mention must be made of the new materials which are now available and which are, to a certain extent, beginning to replace the traditional materials used previously.

Where the soil pipe is concerned, the traditional material has always been cast iron, with the joints 'run in' with lead. Today heavy duty asbestos, pitch fibre and plastics and even copper are all in use, together with a wide range of fittings which greatly facilitate the making of branches for a variety of waste pipes.

New materials have also been introduced for drain pipes, including asbestos, pitch fibre and plastics. The traditional glazed pipes are still used, but it is now realised that a system of pipes which has a degree of flexibility is to be preferred to a more rigid line of pipes.

In view of this a flexible joint has been developed for use with salt-glazed pipes, and there is also an increasing use on sites of pitch fibre and PVC systems. These have the additional advantage of being more simple to join, with the added bonus of being able to undertake pipe-laying in bad weather.

In districts where main sewers are not available it is usually the practice to take the storm-water drainage into soak-aways and the foul water into septic tanks or cesspits.

Storm water is taken from the down pipes and led to soak-aways, which must be the minimum specified distance from the building. The soak-aways consist of pits which are dug at the end of the pipe-run, and filled with clinkers or similar material.

Foul sewage can either be taken to a cesspit or a septic tank, the latter being preferable.

A cesspool is simply a large storage container, and as it does not get rid of the sewage it has to be frequently emptied. The Local Authority usually undertake this work with special vacuum-operated tankers.

A septic tank consists of two chambers, and depends on bacterialogical action for its function.

The first tank is airtight, into which the foul matter enters. The second one consists of a filter bed and a final outlet.

Construction of the receiving tank would be in 9 in. brickwork on a concrete base, and the complete construction rendered waterproof. There should be a cast-concrete top with a standard manhole cover set in it. The filter tank should also be constructed in rendered brickwork on a concrete base. This chamber should be long, narrow and shallow compared with the receiving tank.

Both the inlet and outlet pipes in the receiving tank can be standard, salt-glazed, right-angled junctions, with the long arm down into the tank. The pipe which connects to the filter compartment should be slightly lower than the inlet one.

The filter tank should have a bed of coke and the liquid from the first compartment should be well distributed over it. This can be done by placing corrugated asbestos sheets above the coke bed. The sheets should be placed at a gentle slope and have small holes drilled in them. The liquid can be distributed across the sheets at its point of entry, by means of a length of half-round asbestos gutter placed across the tank. The gutter should also have small holes drilled in it. By means of the asbestos sheet and the guttering the whole of the filter bed is brought into use. The outlet pipe for the filter bed should be at the far end and at the lowest point.

There should be air space above the asbestos sheet and the tank can be covered with loose-fitting concrete slabs.

Specification clauses

The specification clauses appertaining to this chapter are the complete set under Drainlayer, Nos. 25 to 34 inclusive, and clauses 2 and 3 under Excavator.

The fixing of the soil and vent pipe is part of the plumber's work and is covered by clauses 100, 101, 102 and 103.

Quantities of material.

There is very little difficulty in working out the amount of material required for this section of the work. The standard 4 in. dia. salt-glazed pipe is 2 ft. long and the number needed will depend on the actual runs themselves.

Brickwork for the manholes will be in 9 in. work, so every square yard of brickwork will require ninety-six bricks.

Both cement and mortar and tarred gaskin will be required for jointing, and 100 joints take 2½ lb. of gaskin and 1½ cu. ft. of mortar.

The pipes themselves are haunched with weak concrete and a 50 yd. run of pipes requires 5 cu. yd. of concrete. At a 1 : 12 mix this represents approximately twelve bags of cement and 7 yd. of gravel.

Gullies, channels, bends, etc., will have to be considered on site and ordered accordingly.

Costs

It is very difficult to specify definite costs for an item such as drainage, as each case is different, and the cost bears a direct proportion to the length and complexity of the various runs and the number and depth of manholes.

The figures given in the Average Summary are reasonably representative for a separate storm and foul system, with the main sewers within 50 ft. of the building.

The cost of the pipes and the fittings are fairly standard, and the total will depend on the length of the runs.

The bricks for the manholes will probably be available from the main stock, but an extra 1000 have been included.

The largest cost, and the only one on which the reader will have a chance of any saving, is the labour for excavating and laying. This again depends on the length and depth of the excavation and the number of joints to be made, and the cost of this item could rise to as much as £90.

The cost of the connection to the main sewer could also vary, depending on whether the excavation is already made or if a section of road has to be excavated by Local Authority workmen. If the latter is the case, the £15·00 quoted could rise considerably.

Another point which should be considered where costs are concerned is the use of non-traditional pipes. Whilst being slightly dearer per foot run than traditional pipes, there is saving in making the joints both in materials and labour and a saving in the haunching concrete, giving quite a considerable overall saving. There is also no risk of having the cost of cutting out a joint due to a failure when put on test.

Average summary

	COSTS	
	As built	*Using own labour*
	£ P	£ P
4 gullies and grids	5·20	5·20
100 pipes	28·00	28·00
3 manhole covers	12·00	12·00
3 straight channels	·84	·84

Various bends and fittings	6·50	6·50
Gaskin, 5 lb.	·50	·50
1000 bricks for manholes	9·70	9·70
Haunching concrete gravel and manhole gravel, 7 yds.	21·00	21·00
Cement, haunching and jointing, 15 cwt.	8·25	8·25
Labour for excavating and jointing	60·00	—
Connection to Corporation sewers	15·00	15·00
Soil and vent pipes	8·50	8·50
Total	£175·49	£115·49

14. Hot-water cylinder with immersion heater installed (*Courtesy of The British Electrical Development Association*)

15. Drainage detail, showing general run of pipes and a Y junction

16. Drainage details

17. Drainage detail, showing gully and manhole

18. Steps with brick-on-edge risers

19. Fuel-bunker under construction

20. Corner set up for fuel-bunker

Nine

Internal work—woodwork—first fixings— fireplaces—plastering—floor screeds—floor coverings—second fixings

Building Regulation requirements

The main regulation concerned with this stage of the work is that the lower floor shall resist the passage of damp. The necessary measures to ensure this can either be adopted when the sub-floor concrete is laid, or incorporated in the floor finish.

Internal walls and floors of domestic buildings are required to have a specified fire resistance, but with normal building methods this is taken care of.

There are also regulations concerning the thermal insulation of the roof, and this can be met by placing insulating material in or over the ceilings of the first floor rooms (including the bathroom and toilet).

Apart from this there are no direct regulations which apply to work described in this chapter, other than the one referring to the surface spread of flame and those which have been quoted in previous chapters and which overlap into this stage of building. Even so, good building practice should be adhered to throughout.

Information and detail on plans

There is little work carried out in this stage which will be shown and called up in the plans.

The stairs should be indicated on both the ground- and first-

floor plans and sufficient details given so that it can be seen that
they meet the Building Regulation requirements. The position of
stud partitions and clinker block walls should also be indicated.

Floor-boarding will be called up on the section, and the screed
should also be indicated. Internal and external cills should be
shown, and if doors should show on the section they can be
briefly outlined, but the details are not called for.

The main things which should be shown on the plans are the
items which are affected by the Building Regulations.

Constructional details

Once the stage of construction has been reached where the shell
of a house is finished and the roof covering in place and weatherproof, the degree of organisation required to carry the job forward
increases. This is because so many trades have to work either in
conjunction with, or very closely following, one another. A slight
delay with one trade can mean a considerable delay with a following one. If a job is not ready to 'drop on to,' the men concerned
may go to another building that is ready.

Although a separate chapter has been devoted to services, their
installation will have to dovetail in very closely with the work
described here.

Even so, the work described in this chapter offers a lot of scope
for the person doing as much of the work as possible for himself.

When the shell of a building is completed, the first work undertaken is the placing of the first-floor boarding. This is a job which
is well within the scope of the average person, but again, it is one
where assistance can also be given to a carpenter and so slightly
cut costs.

The boards should be tongued and grooved, and are usually
$\frac{7}{8}$ in. or 1 in. thickness. They are fastened to joists with 2 in. lost-
head nails. These are a type of nail, the head of which can easily
be punched below the surface. Widths of boards should be no less
than 4 in. and not greater than 7 in.

The end joints, or heading joints, as they are called, should be well distributed over the floor, and the break should always be on a joist. Heading joints should not be closer than 12 in. to one another.

If the work is being done without using a carpenter, it will be necessary to hire or borrow a set of 'dogs.' This is a tool which enables the boards to be cramped securely together before nailing to the joists. The method of using them is to nail a length of board in position, and then cut and place other lengths until the dogs can be used against an adjacent joist to cramp them tight, prior to nailing. If it is necessary to change the direction of the laying of the boards, this should take place in a doorway if possible.

Some traps will probably be required to allow access to electric wiring, junction boxes, etc. These traps are short portions of board which have the tongue removed. The tongue is also removed locally from the adjacent board. The trap should be screwed in place and so is easily removable to allow access. The electrician should be consulted as to the requirements in this respect, and this is one of the examples of the dovetailing of the work of the various trades. Careful planning over items such as these can save time and trouble later on.

The laying of the floor-boards is an operation that should proceed in parallel with the installation of the staircase.

The main parts of a staircase are the strings, or side members, and the horizontal treads and vertical risers. Newel posts support the banister rails, and in certain instances go straight down to the ground floor to support a landing. A landing which is as wide as the stairs and approximately the same depth is known as a quarter-space landing, whilst one which is twice the width of the flight of stairs is known as a half-space landing. The former is used if it is necessary to run a flight of stairs through 90 degrees and the latter if the flight is to turn back on itself through 180 degrees.

The actual proportions of the stairs themselves depend upon site measurements, bearing in mind the requirements of the Building Regulations concerning the proportions of treads and risers. The treads usually extend over the face of the riser by about 1 in., and this is known as the nosing. As a refinement a scotia or concave moulding is sometimes added under the nosing.

There are two types of staircase which may be considered, one with closed strings and the other with open strings.

The majority of houses have staircases which are of the closed string type. This means that both the treads and the risers are fixed into housings cut in, but not right through, the strings.

The construction of this type of staircase is a specialist's job, as a lot of machining is involved. By the time the shell of the building is completed, this item should already have been put out for quotation, the best one selected and the work already in hand when the first floor-boards are being laid.

In the open string type of staircase, the treads and risers are not housed into the strings, but the latter are cut so that the treads and risers fit on them.

A flight of stairs with open strings is not so complicated as one with closed strings, and the side away from the wall can be built up with a brick or clinker-block balustrade. This balustrade would then be plastered, and the top covered with a wooden capping. This kind of staircase needs intermediate support for the treads, which is given by a carriage. This member is inclined at the same angle as the strings and is midway between them. Blocks of wood, known as rough brackets, are fastened to the carriage, mounted vertically, with the tops coming under, and giving support to, the treads.

The construction of this design of staircase, especially if a straight flight were concerned, would be a feasible proposition for the reader intent on saving as much as possible, providing care was taken in the construction and assembly. The timber for the treads could be bought with the rebate to take the risers already machined in the same way that the fascia board is obtained with the groove already in for the soffit.

The rest of the construction would then consist of some very careful work to ensure that the strings were accurately cut, and the whole staircase assembled correctly.

The staircase is installed by packing so that the bottom riser is in the correct position to match up to the screed when it is laid at a later date. Any parts of the staircase such as the bottom riser or newel-post bases which would actually come into contact with the screed when it is wet, can be protected by pieces of building paper.

As soon as the staircase is in position, both the treads and the risers should be protected against rough usage by nailing pieces of plywood to them.

With both staircase in and floor-boards down, first-floor partitions are the next consideration.

These are necessary when partitions are required on the first floor without any brick walls underneath to carry them. These partitions are usually of clinker-block construction, or studding, or sometimes both are used.

When clinker blocks are used and if they are running the same way as the floor joists, two joists are bolted together to take the load. (*See* Chapter 5.)

Clinker-block walls are sometimes built directly on to the floor-boards, but it is better practice to build them on a timber base which has been rebated out to take the blocks. This method ensures a proper restraint for the base of the wall.

This timber base should be spiked to the double joists if it runs with them, or spiked to each joist if it crosses laterally.

The ends of the walls, where they meet the main walls, will be bonded in by indents previously left. Door and other openings are formed in the normal manner.

Clinker-block walls are generally on the first floor, but if any non-load-bearing partitions are required on the ground floor, they are built directly off the site concrete, providing the area is not too great.

Lintels over doorways in first-floor clinker-block partitions can be a piece of timber, as there is very little weight to go above them.

For any such walls that form a fairly large area on the ground floor, the floor concrete should have been thickened up locally to take this extra load. Small areas, such as larder walls, can be built directly on to the concrete without any extra thickness being required.

Another form of partitioning which is used is studding and plasterboard. A typical application of this is where a partition is required on the first floor to form the boundary of a room, part of which runs over the stair well, and requires a bulkhead. The bulkhead is a means by which a wall can be taken out over a stair well and give local headroom to people using the staircase. The bulkhead slopes back into the room and is covered by lengths of tongue-and-groove floor-boards.

The studding for this partitioning is usually 3 in. × 2 in. roughsawn carcasing timber. This is framed up with due allowances being made for door openings and size of plasterboard panels, and is then covered on both sides with the plasterboard.

Before leaving the subject of first-floor partitions, mention must be made of the Gyproc 2 in. Solid Partitions and the Paramount Dry Partition.

The Gyproc 2 in. solid partition, which is not load-bearing, consists of a core of $\frac{3}{4}$ in. Gyproc Plank, in 2 ft. widths, which is plastered with three coats on each side to give an overall thickness of 2 in.

Paramount Dry Partition consists of two plasterboards separated by, and bonded to, a cellular core, and the $2\frac{1}{4}$ in. and $2\frac{1}{2}$ in. units may be used to construct single leaf partitions. The Partition may be obtained with a surface finish suitable for immediate decoration, or for the application of plaster. Available widths are 2 ft., 3 ft. and 4 ft., while lengths vary from 6 ft. to 12 ft. The honeycomb in the dry partitions is set back from the edge of the board, and joining the pieces is done, broadly speaking, by inserting battens to fit in the rebate of each of the adjacent panels.

There are two further items of work which are done at this stage. These are the installation of the door linings and window-cill boards, normally included in the 'first fixing.' Both are jobs which can be undertaken by the reader.

The door linings are ordered an inch wider than the brickwork in which they stand. In other words, for a lining in a $4\frac{1}{2}$ in. wall the finished width should be $5\frac{1}{2}$ in. This allows for $\frac{1}{2}$ in. of plaster on the face of either wall. Where linings are required in clinker-block walls and stud partitions, the finished width is usually less, and this should be taken into account when ordering. It is very important that the width is stated as finished width, for this means that the lining will be cut from wood which is slightly wider to allow for machining. If the sizes are stated as nominal, then the machining makes them slightly less. For most applications this fact does not matter, but it is important when a full $\frac{1}{2}$ in. is required either side of the wall to accommodate plaster.

Door linings are carefully set up to be true and then nailed to the plugs already built into the brickwork. In the case of linings in clinker-block walls, the linings are nailed direct to the blocks with tapered wire-cut nails, and linings in stud partitions are nailed direct to the vertical members forming the door-openings. In the latter case the linings should have the extra width to allow for plasterboard and one coat of plaster on either side. This again is $\frac{1}{2}$ in. extra for either side.

The internal cills of most modern houses are generally wood, with possibly the exception of the bathroom, kitchen and larder.

Before the internal wood cills are fitted, there is a certain amount of preparatory work to be done. It will be remembered that the cills were left one course of brickwork down at each of the window-openings, and that the frames were supported on bricks across the cavity. When the time arrives to finish the cills these bricks are removed.

Dealing with the external cills first, it must be decided what treatment is going to be used in their construction.

Generally, the modern house or bungalow has tile cills, usually Blue Broseleys. These are clay tiles, approximately $10\frac{1}{2}$ in. $\times$ $6\frac{3}{4}$ in., with three nibs along one short edge.

Tiles are cut so that when they are placed on the slope formed by tucking them under the window-frame, and resting them on the top edge of the outside course of bricks, they project forward about $1\frac{3}{4}$ in., measured horizontally from the brickwork face.

Two courses of tiles are set in place to form the cill, both of them being bedded with cement mortar.

The tiles for the bottom course are cut and bedded so that the nibs are left on and project downwards when the tile is in position. For the second course, the nibs are cut off, so that a plain edge shows above the nibs of the lower course. When the cills are in place the joints between the individual tiles and the two courses of tiles are carefully pointed up with a trowel.

A rather pleasing variation is to make the top course a row of sand-faced tiles that match the colour of the roof tiles. This can easily be done by ordering a few extra under-eaves tiles when the order is placed for the roof tiles.

Another form of cill which is quite common for the average size house is one formed by bricks on edge. Again, these are bedded with cement mortar and pointed up between the bricks. This type, however, does tend to give rather a heavy appearance to the house.

A further type which is sometimes used is the pre-cast concrete cill faced with a reconstructed stone. Provided that there is a local firm which does this sort of work, this kind of cill does not run out as dear as would be expected, and the overall effect is very pleasing. Like the tile cills, the pre-cast ones are fitted at a late stage.

Tiled cills can be fitted after the scaffolding has been taken down, access being gained by opening the windows and working from the inside. Small cills such as toilet and pantry windows can be fitted from a ladder.

Although it may seem foolish to take the scaffolding down and then start fitting cills, this method does ensure that the tiles will not be damaged either by having bricks, etc., dropped on them, or when the scaffold is dismantled.

In the case of a house, however, this procedure may have to be modified if heavy pre-cast cills are being fitted.

For the internal cills, some wooden blocks will be required, $4\frac{1}{2}$ in. wide, to go across the course of bricks, and about $3\frac{1}{4}$ in. high. The blocks should also be tapered, from the bottom to the top, being about 2 in. wide at the base and 1 in. wide at the top.

These blocks are then held in place by a split course of bricks. The bricks are split lengthways, so that the brick is about $\frac{3}{4}$ in. thinner than normal. One end is also cut on a taper to the same angle as the wooden blocks.

When the split course is complete across the window, the blocks should be spaced about every two bricks, and each block will be retained by the tapers of the bricks on either side.

The wood for these internal cills is usually Parana Pine. When ordering, care should be taken to include the projection in the length of each board. The normal thickness is about 1 in.

When setting the boards in position, the tops of the wood blocks are lined through and levelled off with a chisel to suit the thickness of the board and the window-frame.

The projection is cut at each end, the cill pushed into place, and then screwed through to the wooden blocks.

The cills of kitchens and larders are usually tiled, whilst bathroom cills can either be tiled or in Vitrolite. In these cases it is usual to cast a concrete cill first of all, which is low enough to allow the subsequent bedding of tiles or Vitrolite. Such a concrete cill is easily cast in position by blinding the cavity with pieces of slate or tile and then putting a piece of timber across the window opening to act as shuttering until the concrete has set. The tiles or Vitrolite are then set on top of the concrete, the former with cement mortar and the latter with a special form of putty.

Another item which is better in place before the plastering is started is the fireplace. Tiled fireplaces are heavy to manhandle and also very easily damaged. They should therefore only be delivered to the site prior to their installation. When they are in storage against the walls of the room in which they are to be fitted, they should be packed up off the floor to ensure that the weight does not come on any of the tiles. There should also be packing between the top edge of the fireplace and the wall against which it rests.

The first portion to be installed is the burr or fire-back. This is placed in position in the opening of the chimney breast so that it comes far enough forward to be just clear of the back of the opening in the surround. It should also be central in that opening. The space behind the burr is then filled with rubble and lime mortar. The purpose of using lime mortar is to make a more flexible setting, due to the fact that there is considerable heat involved.

As an additional aid to expansion, a piece of corrugated cardboard is placed immediately behind the vertical portion of the burr before the rubble and lime mortar are placed in position. When the fireplace is in use, the corrugated cardboard chars away, leaving a gap between the burr and the filling.

The fireplaces themselves have two metal lugs on either side with a screw-hole in each so that the complete unit can be fastened to the wall.

The surround should be placed against the chimney-breast, packed as necessary, and the position of the screw-holes in the lugs marked on the wall. The fireplace should then be moved away and the wall drilled and plugged for the fixing screws. Wooden plugs should not be used and the drilling should be in the brickwork and not in a joint. Adequate sized fixing screws should be used.

The surround is then moved back into position, an asbestos rope expansion joint added round the opening, and the unit then screwed home.

To set the hearth a bed of lime mortar is spread over the concrete of the solid floor, which will have been thickened up in this area to form a constructional hearth in compliance with the Building Regulations. The fireplace hearth is bedded on this and checked carefully to ensure that it is level.

Free-standing independent domestic boilers also need careful installation and, like fireplace surrounds for Class I appliances, it must be ensured that the appropriate sections of the Building Regulations are met regarding constructional hearths and the permissible proximity of combustible materials. These points also need checking when gas fire units are being fixed and, although there is no specific mention of wall-mounted electric fires in the Building Regulations, care should be taken to ensure that no hazards result from the installation of these appliances.

With all the foregoing items taken care of, attention must now turn to the major job of plastering. Obviously, the reader will know his feeling as to his capabilities of tackling this work. There is no halfway mark. Either a person has enough plastering experience to make him confident (and few outside the building trade will have that experience), or, apart from putting up the plasterboard for ceilings, he will leave the subject alone.

Putting up the plasterboards does not present much difficulty, but the plasterer who is going to do your work for you may either welcome such action or prefer to do the complete job himself.

The fixing of plasterboard is reasonably easy, but there are one or two simple rules which should be followed to give the best results, and minimise the possibility of the subsequent cracking of the ceilings.

There are three types of plasterboard. Gypsum plaster base-board consists of a gypsum core inside a covering of fibrous paper. It is suitable for both ceilings and stud partitions. The board is manufactured in sheets, $\frac{3}{8}$ in. thick and 3 ft. wide. The lengths available are 32 in., 48 in. and 54 in., and the boards are suitable for joist centres of 16 in. and 18 in., spanning either two or three joists.

Plaster lath also has a gypsum core enclosed in specially pre-pared paper. The long bound edges of the laths are rounded to obviate the use of scrim. This is manufactured $\frac{3}{8}$ in. and $\frac{1}{2}$ in. thick, 16 in. wide, and in lengths of 4 ft. and 4 ft. 6 in., suitable for joist centres of 16 in. and 18 in. The choice of either plaster lath or plaster baseboard is mainly a matter of personal preference.

Plaster wallboard has an aerated gypsum core encased in a strong, paper liner, with one side ready for direct decoration. It is manufactured in two thicknesses, $\frac{3}{8}$ in. and $\frac{1}{2}$ in. Widths are 2 ft., 3 ft. and 4 ft., and the available lengths vary from 6 ft. to 12 ft.

It is inevitable that some cutting will be necessary when covering ceilings and stud partitions. All forms of plasterboard are quite easily cut, either by using a saw or by scoring deeply with a sharp knife on both sides and breaking over a straight edge.

The nails used to attach the boards must be galvanised or sheradised, $1\frac{1}{4}$ in. long and 12 S.W.G. for plaster lath and plaster baseboard, and 14 S.W.G. for plaster wallboard. Nails should be at 6 in. centres, and although the nails are driven home, the paper surface of the boards should not be ruptured. Nails should not be nearer than $\frac{3}{8}$ in. to any bound edge, or $\frac{1}{2}$ in. to any edge that has been cut.

It is essential that all cross joints should be staggered and to support the plasterboard at the ceiling edges it is a good idea to insert wood 'noggins' between the joists.

With all the plasterboards in place, the plastering can commence. In order to plaster the ceilings, which is the first operation, it will be necessary to have some planking and either boxes or oil-drums to support them. There should be enough planking and supports to cope with the longest room. The scaffolding should be close-boarded so that the plasterer can reach the complete ceiling without having to get off the scaffolding.

The first operation in plastering a ceiling or stud partition faced with Baseboard is to cover over the joints with scrim. This is supplied in rolls $3\frac{1}{2}$ in. wide. It is manufactured from jute, and has a very coarse and open weave. Neat plaster is mixed up, and the scrim is set along the joints and the angle between wall and ceiling.

Where plaster lath is used, only the external angles and the angles between walls and ceilings need the application of scrim. The joints between the laths themselves are filled with neat plaster. When all the scrimming and filling is set, but not dry, the plastering of the ceiling can take place. This can either be one- or two-coat work.

Whilst two-coat work with its overall thickness of $\frac{3}{8}$ in. gives the ideal job, the majority of ceilings for domestic buildings are in one-coat work. In this case the thickness should not be less than $\frac{3}{16}$ in.

The plaster used is known as Thistle Plaster, or Pink Thistle, but there are many different types of plaster used in building and it is essential to use the correct one.

The plasters we are concerned with here are known as Retarded Hemi-hydrate Gypsum Plasters, class 'B.' Thistle is actually a trade name and the colour can be grey or pink!

The actual plaster specified for one-coat work should be Board Finish Plaster. As its name implies, it is manufactured for use on the various plasterboards.

This plaster is mixed neat, and the correct method is to mix it in a clean galvanised pail, adding the plaster to the water. Cleanliness is essential, and water which has been used for washing plastering tools should never be used for a gauging, which is the name given to a plaster mix.

Lime should never be used when board finish plaster is being applied directly to plasterboards or plaster laths.

With the ceilings and stud partitions completed, attention will now be turned to the walls.

It is usual in domestic buildings to put an undercoat (or floating coat) of plaster and sand, followed by a finishing coat of finish plaster. The floating coat should be $\frac{3}{8}$ in. thick and the finishing coat $\frac{1}{8}$ in. thick. Both these plasters used are again special ones for the particular job in hand. Sometimes the floating coat is made up from a mixture of cement, lime and sand. The use of these materials is not entirely satisfactory, however, due to the shrinkage and subsequent cracking. The gaps that occur between ceiling and wall are usually caused in this manner, whilst the correct application of the suitable gypsum plaster usually eliminates this trouble.

The floating coat consists of gypsum browning plaster, gauged with clean sharp sand. On brick walls the proportion can be one part plaster to two or three parts sand. (The manufacturers' recommendations should be followed on this point.) On clinker block walls or concrete brick walls the proportion should be slightly reduced. Once again the makers' instructions should be followed.

To ensure that the floating coat goes on true, the plasterer uses a rule, or a wooden straight-edge.

Plaster screeds (about 6 in. wide) are set up. The plasterer ensures that these are true in thickness and plumb, and he then fills in the intervening area, using the screed as a guide. A similar technique is used on external corners, such as window reveals and chimney-breasts. Wood battens are set up to ensure a square corner.

When the floating coat is completed, the surface should be well scratched to form a key for the finishing coat.

The finishing coat consists of gypsum wall finish plaster, applied neat with a trowel to a thickness of $\frac{1}{8}$ in. This finishing coat should then be flush with door linings, switch-boxes, etc.

The surface is trowelled to a smooth finish, but should not be over polished.

Although the application of plaster to brick walls and to plaster-boards on ceilings and stud partitions is the traditional 'wet' method of giving a surface finish to the interior of a house or bungalow, it is possible to use a completely 'dry' finish with plasterboard dry linings.

When this method of finishing internal surfaces is adopted, taper edged plasterboards (which have an ivory paper face for direct decoration) are attached to the walls by either dabs of plaster alone, or by a combination of dots and dabs.

With the first method the dabs of plaster, which are about 4 in. diameter and 1 in. thick, are applied to the wall in rows 12 in. apart. The plasterboards are then pressed to these and slid up until the top edge is against the ceiling. Simple lifting wedges are used at the base of each board to keep it in place until the plaster has set.

When the combined dot and dab method is used, the dots are used to 'straighten' the wall, and in this instance are placed on the wall before the main dabs. The dots themselves consist of small pads of bitumen impregnated fibreboard which are bedded to the wall with Board Finish Plaster. They are then lined through and the plaster allowed to set hard.

Plaster dabs are then applied between, and proud of, the dots, so that the individual plasterboards may be bedded to them. A lifting wedge is used to get the boards hard up to the ceiling, and they are then nailed to the dots with double headed nails. These nails are not used as a permanent fixing, but simply hold the plasterboards in place during the critical setting period, after which they are removed. This nailing technique also means that only one lifting wedge is required, whereas several are needed if the boards are fixed with dabs of plaster alone.

Once the dry linings have been fixed the joints are the next feature to receive attention. The depression formed by adjacent tapers is filled with a special compound and a strip of jointing tape is bedded into it. This is followed by a further application of filler to bring the joint flush with the surface of the board. A jointing sponge is then used to smooth the joint and to remove surplus filler.

When this filling has set hard a joint finishing compound is applied in 8 in.–10 in. bands. This is again smoothed and feathered out with the jointing sponge, and this same tool is used for the final operation, the application of a thin slurry of joint finish compound to the entire board surface. This serves the purpose of evening up the texture between the jointed areas and the rest of the plasterboard surface.

The jointing techniques for plasterboards with cut edges, and for internal and external angles is similar to the procedure already described, but in all cases the manufacturer's instructions should be followed carefully.

The same method of joint finishing may also be applied to ceilings, and thus the complete internal surfacing of a house may be done 'dry'. If, on the other hand, traditional plaster ceilings are required, this work should be carried out before the dry linings are attached to the walls.

Once the plastering or dry lining has been completed, attention can then be turned to the floor screed.

If all the ground-floor covering is to be the same, then the screed will be a constant thickness. If, on the other hand, the floor covering is to be mixed, as in the case of thermo-plastic tiles and wood blocks, then the thickness of the screed must be adjusted, to give a level finished floor. In no case should the finished thickness be less than 1 in.

Laying the screed is usually plasterers' work. Once again this is a job about which the reader will have views on his capabilities of undertaking. Whilst not such a difficult task as plastering, a lot depends on the finished results.

The screed must be true and level, it must not crack, and it must also bond completely to the sub-floor. The laying of the actual screed is reasonably straightforward, but the trowelling-up to give the correct finish is more difficult, the art being in 'catching the set.'

It is a Building Regulation requirement that solid floors should be water-proof. If this has not been taken care of by means of a PVC membrane under the sub-floor then provision must be made for protection against damp to be incorporated between the sub-floor and the screed. This can be done by coating the sub-floor (after it has been well brushed) with Synthaprufe and then sprinkling fine gravel over it. The latter forms a good key for the screed.

When using this form of protection against damp care must be taken to see that the waterprooofing compound is taken up the walls to the level of the main damp course.

If it is decided to have a mat-well, this must be allowed for, and the necessary shuttering constructed at this stage. The division between the screed and the front and back steps should also be inserted now. This consists of a metal strip, about $\frac{1}{8}$ in. thick, placed so that it comes halfway under the door concerned. It should be positioned so that the top of it is level with the finished floor covering.

Screeds are best laid by setting wood battens on a mortar bed, so that they are true and level, and their tops represent the finished floor level. The battens are set up to divide the floor into manageable bays, the intervening space is filled in with the material being used, and is then ruled true with a straight-edge, using the screeding battens as a guide.

Another method is to set wooden pads or dots, joining them up by bands of screeds, and then filling in and ruling true, this method being very similar to that used for floating the walls. To facilitate the placing of the screed, the finished level can be marked on the walls surrounding the area concerned.

The mix for the screed should be fairly dry to ensure that there is no cracking due to shrinkage.

Most types of floor covering require a steel trowel finish, but a wooden float finish is more suitable for wood blocks.

When the screed has been put down, it should be allowed ample time to dry. This will usually take about three weeks, but will mainly depend on the weather conditions prevailing at the time.

For the covering of the floors themselves there is a wide choice of materials facing the reader, and a wealth of literature on the subject is available from manufacturers.

One popular floor covering is the 9 in. square thermoplastic tile, which can be obtained in a number of attractive designs. These can be ordered on a supply and fix basis, and there is also a range of thermoplastic tiles available which are suitable for the home enthusiast and which do not need specialist labour for fixing. Attractive features can be made of these floors by the addition of coved base skirtings, cut-out tiles and feature and edging strips.

Whilst some form of thermoplastic tile is the more conventional form of floor covering for modern domestic buildings, linoleum tiles could also be given serious consideration.

They are laid on a good screed and will take a good deal of wear, especially in a house where foot traffic is slight. Under these conditions their life should be between ten and fifteen years. There may be areas such as thresholds and where people stand at sinks, where the tiles may need replacing within the ten years, and it would be a good idea to obtain a few extra tiles, so that the tiles in these areas could be replaced if necessary.

Of course, these areas will also be the wear spots for thermoplastic tiles or any other type of floor covering.

From the foregoing it will be seen that the reader intent on cutting costs could very easily lay his own floors.

The procedure for laying both linoleum and thermoplastic tiles is similar, and the manufacturers of both issue descriptive leaflets which detail fully the correct way in which to set about the job.

Wood blocks are another popular floor covering, but as mentioned previously they are thicker than thermoplastic or lino tiles and the thickness of the screed must be adjusted accordingly.

There are various types of wood used for the blocks, and care should be taken to ensure that the one chosen has a good hardness factor.

The blocks can be laid to several designs, and probably one of the most popular is the single herringbone. The simplest is the brick pattern, in which the blocks are laid end to end, the joints in adjacent lines being broken.

When laying this type of floor, the marking out procedure is similar to that required for the lino or thermoplastic tiles. That is, the centre of the walls and the centre of the room needs to be known, and marked before laying commences.

Allowances must be made for expansion, and it is usual to insert

F

a thin strip of cork round the edge of the block floor to accommodate any movement. A suitable adhesive to use is 'Synthaprufe.'

Another form of flooring which is worth considering is cork tiling. These are more resilient and warmer to the touch and may be preferred in a bedroom. In overall effect they give a pleasing neutral colour. Once again, 'Synthaprufe' can be used as the adhesive. In addition, this type of tile should be fastened at each corner with steel pins. The pins are hard enough to penetrate the floor screed.

As soon as the flooring has been completed, the 'second fixings' can be commenced. This is certainly a job which the reader should be able to undertake for himself. The second fixings include fixing architraves, hanging all doors and putting the stops on the linings, fitting all door furniture, larder and cupboard shelves and fixing the skirting-boards.

The architraves are the pieces of wood which cover the joint between the door lining and the plaster, and are required on both sides of the wall in which the lining sits. They are cut from strips of wood approx. 2 in. × ¾ in. They can either be rounded on both edges or square on one edge, with a flat on the face which tapers down to a thinner section on the other edge. The former type, if anything, has the more pleasing effect.

The architraves are nailed to the door linings, and the two vertical ones should be carefully mitred to the pieces across the top of the door. Architraves should go down to floor level and must be carefully cut to match in with the skirting.

If rounded architrave is used, together with skirting-board which is also rounded on the top edge, the method of joining is as follows. A top corner of the skirting is cut off at 45 degrees. The bottom of the architrave is also cut at 45 degrees from the outside edge and towards the inside (or door) edge, but stopping at the start of the round portion. A vertical cut is then made from the bottom of the architrave to meet this cut. When skirting and architrave are assembled in their correct positions, the outside (or wall) edge of the latter will meet the top edge of the former, and the end of the skirting will be finished off by the rounded

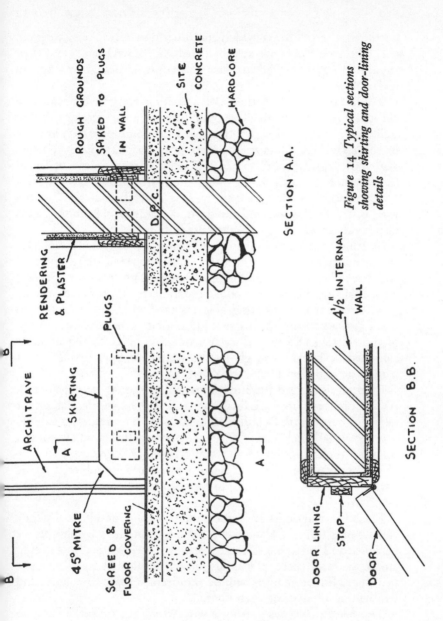

ARCHITRAVE

ROUGH GROUNDS SPIKED TO PLUGS IN WALL

RENDERING & PLASTER

PLUGS

SKIRTING

A

A

45° MITRE

SCREED & FLOOR COVERING

B

B

SITE CONCRETE

HARDCORE

D.P.C.

SECTION A.A.

4½" INTERNAL WALL

DOOR LINING

STOP

DOOR

SECTION B.B.

Figure 14 Typical sections showing skirting and door-lining details

portion of the architrave which continues down from the 45 degree cut. The architraves are attached with $1\frac{1}{2}$ in. oval nails, and they should be lightly tacked in place and adjusted before driving the nails fully home.

There are several methods of attaching skirting-boards. The cheapest method, which is quite common in modern house building, is to nail or screw the board to the plugs which are put in the raked-out joints in the wall prior to plastering. A better way, but slightly more expensive, is to spike a piece of wood (known as a ground) to the plugs, and then fixing the skirting-board to that. In that case the plugs would be cut back to the level of the brickwork, and the ground would equal the thickness of the plaster which would come down to it.

Skirting-board heights are something which should be considered carefully. Some modern houses have a skirting-board which is only $2\frac{1}{2}$ in. or 3 in. high. Although slightly more expensive a 4 in. skirting does give a room a much better appearance.

To ensure that the skirting-board is hard down to the floor when it is fixed, a kneeling-board should be used. This is placed on top of the skirting and sloped away from this point to the floor. It can then be knelt on to give some weight to the skirting whilst it is being nailed in place.

Internal doors are the next item which should be considered. All the modern doors are flush-faced, and the covering can either be ply or hardboard. Standard door sizes are 6 ft. 6 in. × 2 ft. 6 in. and the usual thickness is $1\frac{5}{8}$ in. It may be that a small door is required for the larder or the airing cupboard, but even so, one of the standard sizes of smaller doors should be adhered to. If not, the door will have to be purpose made and consequently more expensive.

Flush internal doors are not completely solid, and basically they consist of a frame which is covered on both sides by the ply or hardboard. The frame is then blocked where hinges, locks, etc., are to go, and these points are clearly indicated by distinctive marking. Hinges or butts will be required to hang the doors, and two will be needed for each door.

There are two types of hinges with which the reader is likely to be concerned. One is the standard hinge and is usually 3 in. long for internal doors.

The other kind is known as 'rising butts,' and are designed to enable the door to rise as it opens in order to clear a carpet. This type of hinge is only successful if the edge of the carpet is a fair way into the room. If the carpet is close to the door, rising butts do not allow it to be raised enough to clear the carpet in time, as the maximum lift occurs only when the door is fully open. Rising butts do have the advantage, however, in that they make a door self-closing.

Where a carpet is involved, a better method is to put down a threshold strip under the door thick enough to allow the door to be above the carpet when the former is fitted. The threshold strip should be well bevelled on both sides, and can be screwed to the floor, using rawlplugs and counter-sunk screws.

When a door is hung there is a certain amount of trimming necessary, and the top and bottom of the door should be carefully planned to give the required clearance. This should be the absolute minimum, as eventually the door will shrink and any gaps will be increased.

If, for any reason, it is required to cut a strip off a door, using a saw, care should be taken to see that this is done equally from top and bottom. If not, the bottom rail of the door may be completely cut away, exposing the hollow structure. If this happens, the resulting tunes played by the natural draught can be annoying to say the least.

With doors in place, the stops can be 'planted on' the door linings. The stops are pieces of wood, $1\frac{1}{2}$ in. $\times$ $\frac{1}{2}$ in., nailed to the linings and serving the same purpose, in a cheaper manner, that the rebate serves in the external frames.

Parallel with the fixing of the stops can go the fixing of the door furniture. There is a very wide choice of fittings available. Probably the most satisfactory is the plastic lever mortice latch furniture, used with a mortice latch and striking plate.

A mortice latch is a small unit which fits into a mortice cut in the blocked portion of the door-stile. It is retained by two counter-sunk screws and the retaining plate itself is also let into the stile of the door. The latch is then operated by a square spindle which fits into the plastic levers which are spring loaded and screwed to both faces of the door. The assembly is completed by screwing a striking plate to the door lining. The plate is set flush in the lining, and part of the lining behind it must be cut away to house the latch when the door is closed.

A mortice latch is non-locking, but there is a variation which has a locking snick. With this the doors can be closed, and the operation of the snick on the hall or landing side of the door means that anyone who has gained unauthorised entry into the room is unable to go any further.

The front and back doors require slightly different treatment. They are usually wider than internal doors, being either 2 ft. 8 in. or 2 ft. 9 in., and the thickness is $1\frac{3}{4}$ in. or $1\frac{7}{8}$ in. They are hung with 4 in. butts.

For the front door, there is the choice between a rim lock and a mortice lock. A mortice lock is fixed into the stile of the door in the same manner as a mortice latch, but a rim lock is fixed by screws to the face of the door. If a rim lock is used it must be the night latch type. That is it is not operated by any handle outside the door but by a small knob on the inside.

A very satisfactory way of locking a front door is to use a mortice sash lock. This has a non-locking latch which is operated by spring-loaded handles on either side of door, and in addition there is a dead lock. The dead lock is not self locking and has to be key operated.

Draw bolts are usually added to the top and bottom of the door as an extra precaution.

A letter-plate will also be required for the front door, and this should be cut into the block indicated on the door.

Similar fittings can be used on the back door, or it may be preferred to fit a combined rim lock and latch. Draw bolts could again be fitted as an added security measure. When both the external doors are fitted it will be necessary for the glazier to return to site to fix the glass panes in these two items.

The fixing of larder and cupboard shelves is a simple matter and needs no explanation. In the airing cupboard wood slats are fitted instead of shelves to allow warm air to circulate.

The trapdoor into the loft will require finishing. This is done by adding a lining to the opening formed by the rough timbers of the ceiling joists. This lining can be cut from scrap pieces of floor-boarding and should be level with the plasterboard ceiling, but should only go about halfway up the depth of the ceiling joists. By stopping at this point it forms a landing for the trapdoor itself, which can either be made from floor-boarding or hardboard, suitably reinforced.

This work is then completed by adding pieces of architrave to cover the joint between the lining and the plaster ceiling.

Finally there is the bath panel to be fixed. This is a sheet of enamelled hardboard, coloured to match a coloured bath or black for a white bath. A simple frame is made up from 2 in. $\times$ 1½ in. wood, with uprights at every 15 in. The bottom of the frame is nailed directly to the floor-boards and the top fits under the rim of the bath. The hardboard is then fixed to the frame, using raised counter-sunk screws. These should have a chromium-plated finish.

At the bottom of the panel, a piece of quadrant is sometimes fitted, or a piece of skirting-board can be used. This is more in keeping with the finish to the walls of the bathroom, and also prevents the hardboard being damaged by kicking.

Specification clauses

There are two remaining clauses in the sample specification which concern the Bricklayer in this part of the work. These are numbers 49 and 62. Clause 59 may also apply for such items as plugs for the internal cills.

Under Joiner the following clauses apply: 69 to 78 inclusive, and 81 to 83 inclusive.

Clauses 121 to 129 inclusive will apply to the Plasterer. Any variations in the use of materials, etc., will, of course, be reflected in the specification.

As mentioned earlier, it will probably be necessary for the

Glazier to return to site, and clause 132 covers the glazing of the external doors.

As well as the foregoing items in the specification, there is a schedule of doors, giving sizes, hinge and furniture details.

Quantities of materials

For the majority of work discussed in this chapter the required quantities can be easily obtained by direct physical measurement. In this respect the ordering is made easier than when quantities of materials have to be extracted from the drawing in order to erect the building.

The first item required is floor-boarding. This is the area of the first floor, less the area of the stairwell. The boards are ordered as a number of squares, a square being 100 sq. ft.

Studding, if required, is rough-sawn material, and the quantity can be estimated by checking the length of the partition concerned, and allowing for full uprights at 18 in. centres and cross members at 2 ft. centres.

The quantities of door linings, and the sizes of cill-boards are simply a matter of a site survey.

Quantities of Broseley tiles for the external cills can be easily calculated from the fact that a tile is $6\frac{1}{2}$ in. wide. It must be remembered to double the number worked out, as two courses of tiles are used. A 6 ft. 6 in. window requires twenty-two tiles, and 150 should be sufficient for the complete house.

The two major items, of course, are the quantities of materials for the plastering and for the floor screed.

As a fair average, the house of approx. 1000 sq. ft. in area has just under 100 sq. yd. of ceiling area and just under 300 sq. yd. of wall area.

Taking the ceilings first, 100 sq. yd. represents 120 plaster base-boards, size 3 ft. × 2 ft. 6 in., 115 size 3 ft. × 2 ft. 8 in. and 75 size 3 ft. × 4 ft.

In terms of plaster lath, this area represents 200 boards size 42 in. × 16 in., 170 boards size 48 in. × 16 in. or 150 boards size 54 in. × 16 in.

For one-coat work 100 sq. yd. of ceiling needs 10 cwt. of board finish plaster, two rolls of scrim and 8 lb. of nails.

When considering the walls, 1 ton of plaster will have approx. 250 to 300 sq. yd. covering capacity, and will need 6 yd. of sand.

For the floor screed, the 500 sq. ft. floor area gives a volume of 1·5 cu. yd. at 1 in. depth. With a 1 : 4 mix, this will require 10 cwt. of cement and 2 cu. yd. of sand.

With the plastering and floor screed completed, the final quantities covering doors, architraves, skirting-boards, shelving and floor covering are again simply a matter of a site check.

Costs

Once the shell of a house is completed, a very careful check needs to be kept on expenditure, as it is very easy to spend a lot of money on fittings and internal finishes.

By doing the carpentry oneself, it is possible to save the cost of the first and second fixings, but there are only two slight savings that can be made in the cost of materials for this work. The first is to obtain the timber for the door linings in lengths and make the linings up on site. If the linings are ordered complete, it can add to the cost. The other item is in the choice of internal doors. Those with ply facing can cost more than those with hardboard facing, and this can mean a difference of several pounds on the complete house.

The largest labour charge for this section of work is for the plastering, and there is little that can be done to reduce this if traditional materials are to be used. The price of £120·00, given as an example, could vary due to the locality and the availability of labour. It may be possible to obtain the services of a jobbing plasterer, and acting as a labourer the reader could perhaps reduce this cost to £60·00 or £80·00.

If a dry finish is contemplated, the materials will probably cost more, but the saving could come on the labour, as the complete ob could be undertaken oneself.

The other main item where saving could be made is in the floor covering. By using materials which can be laid without a specialist a considerable saving could be made.

Average summary

	COSTS	
	As built	Using own labour
	£ p	£ p
4½ sq. 1 in. flooring (T. and G.)	32·50	32·50
Staircase	35·00	35·00
150 ft. of door-lining timber	9·00	9·00
80 ft. of 3 in. × 2 in. studding	3·00	3·00
30 ft. of Parana pine cill board	2·90	2·90
150 Blue Broseley tiles	4·60	4·60
Vitrolite cill	2·00	2·00
Joiner—first fixings	35·00	—
One fireplace, burr and fret	40·00	40·00
One fireplace, burr and fret	25·00	25·00
100 sq. yd. of plaster lath	16·00	16·00
2 rolls of jute scrim	1·50	1·50
8 lb. galvanized nails	1·40	1·40
10 cwt. board finish plaster	6·45	6·45
1 ton Thistle Browning	12·50	12·50
6 yd. sand	12·00	12·00
Labour for plastering and screed	120·00	—
Labour for plastering, excluding rough labour	—	80·00
10 cwt. cement for floor screed	5·50	5·50
2 yd. sand for floor screed	4·00	4·00
Supply and fix thermoplastic tiles to ground floor	60·00	—
Self-lay tiles to ground floor	—	40·00
Joiner—second fixings	30·00	—
300 ft. architraves	5·00	5·00
300 ft. skirting	7·50	7·50

9 plyfaced internal doors	22·50	22·50
2 external doors	10·00	10·00
150 ft. of 1½ in. × ½ in. stops	3·00	3·00
Bath panel hardboard	1·50	1·50
Door furniture, hinges, etc.	20·00	20·00
Roof insulation	15·00	15·00
Sundries, nails, wood, etc.	10·00	10·00
Total	£552·85	£427·85

Ten

The finishing touches—steps—painting—walls and fencing—fuel bunkers—paths—garages— sheds

As soon as the house has reached the stage described in the previous chapter and the Building Inspector is satisfied that it is habitable, occupation can take place. There will, of course, be a variety of jobs still to do, but these can be undertaken at one's leisure.

The priority which is given to these jobs depends both on personal choice and also the time of year in which the move takes place. If it is during the winter there will be an urgent need for garden paths, steps and fuel-bunkers. However, keeping to a logical sequence, there are still two items concerning the house itself which have not yet been dealt with.

Steps are the first item. Generally there will be three, one for the back or side door, one for the french windows, and the main front entrance step.

The steps for the side entrance and the french windows can be quite straightforward. At these points, the brickwork of the outer skin of the cavity wall should have been left down a couple of courses (*see* Chapter 4). In its simplest form the step can be level with the floor surface and also flush with the outside face of the brickwork.

A piece of board can be placed across the outside skin of the brickwork where the step is to come, and the cavity must be 'blinded' with pieces of tile or slate. The mix used should consist of granite chippings (for hardness), sand and cement. The proportions can be 1 : 2 : 2. In the case of the rear or side doorstep, the material used will butt up to the steel dividing strip mentioned in the previous chapter. In the case of metal french windows the material will butt-up to the metal frame.

The shuttering should be removed before the steps have set completely hard, in order that the surface can be brought up to a smooth finish with a trowel. In addition, the sharp edge to the step should be slightly rounded in order to prevent damage when in use.

The front step is likely to be more complicated than those just described, but the actual details will depend on the situation of the house, on the site, its design and personal choice. The actual step at the threshold can be made in the manner just described. In addition, there may be a step under a canopy or porch, and there are a variety of ways of treating this. One method is to lay quarry tiles on a concrete base, and finish off with a granolithic edge.

When constructing these various steps, consideration should be given to the finished path levels, and how these tie in with pavement level, garage drives, etc.

The other major item needed to complete the house is the painting and decorating. In a new house there will be quite a lot of movement due both to the structure settling, and timber shrinkage. In view of this the initial decorating which takes place should only be regarded as fairly temporary. Once the structure has settled down it is then possible to indulge in the use of a wider variety of decorating materials, including wallpaper. In this sphere a tremendous amount of literature is available, both from the manufacturers and the 'Do It Yourself' Press, and it is not proposed to deal at length with this subject.

Probably the best material to use for the first application to the new plaster of the walls and ceilings is an oil-bound distemper. It is a good idea to try and standardise on only a few colours, and if this is done the material can be bought in bulk rather than small individual tins.

Before any of the woodwork is painted it should be carefully rubbed down, all nails punched below the surface, and any necessary filling done. The more care that is taken with this preparatory work, the better the finished results will be. When this has been done, the knots in the woodwork should be sealed by the brush application of knotting varnish. The woodwork should then be given a coat of lead primer paint before the required undercoat and top coat of paint is applied.

The external woodwork should be given the same careful treatment before painting, although certain items such as door-frames and fascia boards will have already been knotted and primed before being built-in. If metal windows are fitted they will only need to be given the required under and top coat of the chosen colour. It is a good idea, however, to seal between the outside of the frame and the brickwork with one of the various sealers now on the market. This material can either be bought in strips, or in tubes with a special squeezing attachment at the end. A narrow fillet of this material is applied all round the frame and this becomes inconspicuous when painted. If asbestos gutters and down pipes are used on the house, a special sealer should be applied before any painting is done. If this is not done, a satisfactory finish will not be obtained by painting directly on to this material.

The construction of the front wall is an item which should certainly be undertaken by the reader. The usual form which this takes is a solid 9 in. wall approx. 2 ft. high. It may be, however, that there is a Vendor's Clause in the agreement which specifies the exact requirements. There is a tendency nowadays on some estates to have open frontage. Obviously any rules such as this will have to be adhered to.

If a brick wall is to be built, the foundations will have to be marked out and dug in exactly the same way as for normal brickwork. The trench will have to be pegged to give the required depth of concrete, but in this case 6 in. depth should be sufficient. The foundations should be enlarged in width where it is necessary to build piers for entrance and garage gates.

The portion of the wall which is below ground level can be in common bricks, changing to facings as the work emerges above ground.

Piers which are 13½ in. wide will have a space in the centre which

should be filled with concrete. Any piers which are going to have
gates hung on them would also benefit from a central reinforcing
bar in the concrete and this should be put in when the foundations
are poured.

Hinge pins for the gates should be built in as the work proceeds.
If this is not possible, portions of brickwork should be left out, so
that they can be built in at a later date.

There are various forms of decorative finish which can be
applied to the tops of the piers and wall. One of the simplest is to
finish off with bricks on edge. In the case of the piers these bricks
can have a double course of Broseley Tiles underneath them,
projecting about 1 in. all round the pier. When setting out the
piers for garage and entrance gates, a careful check should be made
to see that the correct space is left between them. That is, allowance must be made for the projection of the hinges and catches
as well as for the actual gates themselves.

Whilst the wall at the front boundary is most likely to be of
brick, there is still the normal garden fencing to be taken care of.
It will depend on the individual site as to the exact number of
fences, which have to be supplied and maintained by each individual householder.

It is the usual practice these days to use wire and conrete posts
for these boundary fences. Concrete fence-posts are more durable
than wooden ones and the fence itself can be made either from
single strands of wire alone, or from chain-link fencing. If the
latter is used it will still be necessary to have the single strands of
wire, and then chain-link fencing is pulled taut and attached to
them.

If a large number of fence-posts are required, there is no reason
why they should not be made, rather than bought. If only a few
are required it may be possible to join forces with one or two of the
neighbours and so make the project worth while.

There are various widths of chain-link fencing available, and the
posts must be constructed to suit these. A suitable width is 42 in.
and the height of the posts above ground should be 44 in. Approximately 2 ft. would be needed to go into the ground, giving the
overall post size as 5 ft. 8 in.

Above ground line the posts usually taper, and this could be 5 in. square, to 4 in. square at the top. Below ground, the post can be straight at a constant 5 in. square. The top of the post can either be rounded or chamfered.

Corner, or Arris posts, have to be supported to take the strain of the wire when it is pulled taut. To achieve this, these posts are braced by concrete posts at 45 degrees to them, resting in notches in the vertical member. These posts can also be cast by making slight modifications to the mould for the standard posts.

The mould itself can be constructed in 1 in. timber and consists of two sides, securely screwed to a back-board. There will be a join in the sides at the ground line, when the section changes from straight to tapered. The mould can be blocked across at the bottom end with a loose-piece of wood. At the top end it can either be given a chamfer by two pieces of wood, or it can have a half-round insert of thin mild steel plate, to give a rounded top to the post. The side of the mould should be rebated to allow for the thickness of the plate. Two or three wooden straps will be necessary to stop the sides of the mould bowing out when the concrete is poured in. These should be removable and only clip over the sides.

To make the straining posts, the top of the mould can be blocked off by a piece of wood placed on the slant to suit the angle at which these posts will lie. To form the corresponding notch in the Arris post, pieces of wood, suitably shaped, can be screwed to the side of the mould as required.

When casting the posts, reinforcing bars will be required. Two ⅜ in. diameter rods should be sufficient. Loops of wire will also be required to be cast in the posts, and it is through these that the straining wires will pass when the posts are finally erected. These loops can be cut from the rolls of galvanised wire which will have to be bought to use as straining wires. Short lengths should be cut and bent to a loop, and the 'tails' are bent so that they will key into the concrete of the post. Three loops are sufficient for each post. One should be just above the ground line, one near the top of the post and the other centrally between the two.

Great care should be exercised when taking the posts from the mould. Two or three days should be allowed to elapse before attempting to do this, and even longer should be allowed in damp weather. The wooden straps should be removed and the sides of the mould carefully 'sprung,' the mould held up in a vertical position and the post gently eased from it.

When setting up the fence-posts, they are placed approximately 8 ft. to 10 ft. apart. Holes are dug for them to the required depth and about 1 ft. square, the post is placed in and wedged and packed to the correct position and the hole is then filled with concrete. It will be necessary to use a line to ensure that the posts are in the correct position, and each post should be braced until its base concrete has set. The posts are set so that the wires are on the boundary, with the posts on the fence-owner's land.

When all the posts are in position, the straining wires are passed through the loops on the posts, anchored at one end and adjusters are used at the other end to pull the wire taut. The chain-link fencing is then unrolled and fastened to the straining wires with pieces of light-gauge galvanised wire.

If solid fuel appliances have been included in the heating arrangements, then provision will have to be made for fuel storage.

The best form of fuel storage is a brick bunker, and the cost of building this with bricks which are probably already available is negligible compared with the cost of a pre-cast concrete or galvanised steel one of equal capacity.

Siting should be given some consideration. It should be easily accessible both from the point of view of the householder and the fuel supplier. If possible its position should be tied in with any external lights that are fitted.

It may be feasible to add the bunker to the rear of a garage, or on the other hand it may be possible to build it into a bank, or incorporate it in any building up that has to be done, so that only the front and the top are visible.

Coal is approximately 40 cu. ft. to the ton, and coke is just about double this figure. This means that to contain a ton of coal the bunker could be 2 ft. high, with a base of 5 ft. × 4 ft. Keeping the same height, the compartment for coke would have a base of

4 ft. × 8 ft. In actual fact, a more practical height would be 3 ft., and the other dimensions could either be left or modified accordingly, depending on how much coal or coke it was decided to store. The ideal height for tipping a sack of coal is 4 ft. 6 in., but for practical purpose it may be necessary to construct the bunkers nearer to 3 ft. or 3 ft. 6 in. high.

When these points have been decided, a concrete raft should be put down on a hardcore base to embrace the dimensions decided on. This base should be about 4 in. thick.

The setting out and construction of the bunker should not present much difficulty, even for the person who has done little or no work on the house itself. In fact the construction of a fuel-bunker could form a very good introduction to the art of brick-laying.

Before setting up any corners, the outline of the bunker can be marked on the concrete base, and bricks can be put down to check the positions of openings and dividing walls. When this has been satisfactorily worked out, the bottom course can be properly laid with cement mortar. Corners can then be set up and the brickwork proceeded with in the normal manner.

Depending on the design of the storage unit decided upon, it may be necessary to put a concrete lintel across the opening from which the fuel is to be taken. The method of constructing these lintels has been described in Chapter 5.

The roof will require some form of opening to enable the fuel to be tipped, and there are several ways of doing this. One is to have a completely wooden top to the bunker, a portion of which is hinged to allow delivery access. The boarding should be covered with one of the various roofing felts available. As an alternative a portion of the top can be cast in concrete (which would have to be suitably reinforced) and a lift-off access panel could be formed in wood. The actual method used will depend largely on the individual design and size of the bunker, but in any case the roof should be given a slight slope to throw off rain water.

The construction of a path is quite a straightforward item, but the actual siting is something which should be given some consideration. Generally speaking it is normal practice to put a path completely round a house. This has the advantage of giving some measure of protection to the foundations by forming an apron above them.

When constructing the paths which in general go from the house to the bottom of the garden, thought should be given to such things as the final garden lay-out and the position of that very important item, the clothes-line. If it is decided to have the lawn and flower garden full width behind the house, with the kitchen garden behind it, the main garden path would probably be more suitable running down one side. However, if this path is the access path for the clothes-line, care should be taken to see that any clothes on it do not foul the fence. The path down the centre makes the problem of washing easier, but it does split the garden into two definite portions.

When making preparations for laying a path, the vegetable soil should be removed and a firm base of hardcore or clinker put down. Screed-boards are set up above this, and the concrete poured in between them. Long strips of path should be split up with an occasional expansion joint which can be a strip of thick bitumastic roofing felt. A suitable mix for the concrete would be 1 : 6, using an 'all in' gravel.

Paths which are built adjacent to the house should be constructed with a slight fall away from the building. Sometimes it may be desired to construct either dwarf walls or perhaps a lean-to porch or conservatory. In this case the path could be made wide enough to act as a foundation for either of these items. In the case of a conservatory, it would probably be necessary to thicken up the concrete locally where any heavier structure is to be built. Time spent on thought in planning for future work can save time and unnecessary labour later on.

When constructing the paths around the house, protection must be given to the drainage gullys, the tops of which will be approximately level with, or slightly below, the finished path level. The usual way to do this is to put a small concrete kerb or upstand all the way round the gully. If a gully is in such a position that it can serve as a drainage point for the path, then the kerb will have to make allowances for this.

Where an inspection cover comes in a path, its top should be at path level, and the frame of the cover will be retained by the path concrete. The covers should be sealed with a mixture of grease and sand.

Steps are quite likely to be required and these can be made in a variety of ways. The simplest method is to make some shuttering and pour the concrete directly into this. The shuttering would have two side pieces cut like open strings of a staircase, and boards nailed across the front like risers. The shutter-box is then set in position, the concrete poured in and screeded level at each step.

A pleasing alternative is to construct the steps with a "brick on edge' front or riser. If this method is adopted care should be taken to see that the bricks used are hard enough for the purpose. If bricks of insufficient hardness are used they are liable to chip in use and if this happens they become very difficult to replace.

The bricks are set to a line in the form of a soldier course using ordinary cement mortar. The joints between them are pointed up in the normal manner. When the mortar has set, the concrete for the tread can be poured in behind the bricks. The standard length of a brick, approximately 9 in. long, is too great to use in full for the rise of a step. The bricks can either be cut to the required length or else the steps can be so constructed that each soldier course is set a little below the level of the step below it.

A further variation of the soldier course step, is to make a curved front. The method of construction is similar, except that the soldier course is set to a gentle curve. A line will still be needed to keep the bricks level, but the curve can be gauged by eye.

When constructing paths and steps on areas that have been made up, it should be ensured that the ground has settled enough before the work is undertaken. Work on made-up ground should be left as long as possible and it could even be worth considering some form of temporary paving or covering whilst the ground is settling.

We now come to garages and sheds. Garages built directly onto the house have been mentioned earlier in this book, but it is sometimes decided that a garage is to be sited away from the property. If such a garage is built at the same time as the house, then details of its position and construction will have been included on the house plans submitted to the Local Authority before building began, and so its building can follow on as soon a the house is finished.

However, if the garage is to be built some time after the house is occupied, and details were not included on the original plans, then it must be treated as a separate application, and the necessary plans and information submitted to the Local Authority. This also applies to pre-fabricated garage units and car ports.

On the other hand, small sheds, greenhouses and similar garden structures are exempt from constructional requirements if they do not exceed 1,000 cu. ft. in volume and providing that they are situated at least 6 ft. from the house.

The constructional details of sheds do not present very serious problems. It is possible to buy a variety of sheds already made, but it is quite simple to construct such an item oneself.

Such a shed should be sectioned, and the whole structure should be mounted on a firm base. The base should take the bottom of the shed above the earth level. If the shed is to have a wooden floor, the base can consist of a dwarf brick wall with intermediate piers if it is necessary to support the floor, other than round the edges. In other cases, the shed can be mounted on a concrete base.

Various forms of boarding are available. The most common being tongued-grooved weather-boarding. A more pleasing alternative and a more satisfactory one, is to use ship-lap boarding. With this the boards are rebated into one another and the overall effect is more pleasing than the boarding previously mentioned.

Roof covering is usually roofing felt, over a close-boarded base. Another type of covering is to use tiled strips made from a similar material to roofing felt. It may be that there will be a stack of tiles left over from the house construction, and if there is a sufficient quantity, these could also be considered for use. If they are used, the roof structure must be modified to take the increased weight.

Where a garage is concerned, it will be found that many points in its construction are similar to those in a house, and all the principles involved have been described earlier in this book.

The garage will have to be marked out and foundations dug for it, in exactly the same way as those for a house are marked out and dug. Concrete foundations are then put in and the brickwork is brought up to D.P.C. level. The exact decision as to which is the correct level will be influenced by the approach to the garage. When the brickwork is at D.P.C. level, the hardcore and concrete can be put in then, or it can be left until the garage is finished and the floor concrete and drive can be laid at the same time.

When constructing the garage, the frame for the main doors and any side door and windows that may be required, are set up in exactly the same way as the frames for the house were set up and built in.

For the roof, purlins and corrugated sheets are usually used. These slope from front to back and the sheets are hidden from front view by a low brick parapet built above the main door-frame. As an alternative, a pitched roof may be built and covered with tiles and ridge to match the house. This type of construction requires the attendant woodwork in the form of rafters, ridge-board, etc.

If it is decided to have a garage at the same time as the house is built, it is worth while considering building it before work on the house is actually started. Doing this obviates the need for a shed, and would give a dry lock-up storage for building materials. Obviously the siting of the garage will have some bearing on the decision taken, as for example, it would not be possible to do this if the garage is to be built immediately on to the side of the house.

Specification clauses

The only clauses left to mention in the sample specification are those which deal with the internal and external decorating. These are the final ones, Nos. 133–142 inclusive.

Quantities of materials

Once the final stages of building are reached, it becomes difficult to give an overall estimate for such items as paths and steps. Invariably there will be materials left over from the major building operation, and, when these are exhausted, cement and gravel, etc., can be bought as required. Initially, when there are all the paths to lay, and fuel-bunkers to build, sand and gravel can be bought in larger quantities to continue to obtain the price concession. As the work to do becomes less, so the quantities can be reduced.

A 9 in. wall on a 40 ft. frontage and 2 ft. high will need in the region of 2000 bricks, and a garage a further 2000 to 2500. These should, of course, be included in the main brick order.

The garage and drive will probably use $1\frac{1}{2}$ tons of cement, 9 to 10 yd. of gravel and about 1 yd. of sand for bricklaying. The cement for bricklaying is included in the $1\frac{1}{2}$ tons.

Paths could account for a further ton of cement and another 6 yd. of gravel, whilst the front wall, depending, of course, on its length, could use $\frac{1}{2}$ ton of cement and 1 yd. of both sand and gravel.

Turning to inside work, distemper for the wall and ceiling finishes can be bought in 14 lb., 28 lb. or 56 lb. packages, as well as in smaller 7 lb. tins. Average covering capacity is 45 sq. yd. for 7 lb. of oil-bound water paint, one coat only.

The ceiling area is approximately 100 sq. yd. so this should require one 28 lb. packet for two coats.

The wall area is in the region of 300 sq. yd. so the requirement will be 90 to 100 lb. of material.

Costs

By the time these final stages of building are reached, the reader will have a very good idea of the prices of the various materials.

If a house is being built by a builder, it will be finished off completely before occupation. When building by employing direct labour or by doing the work oneself, many people pay for the work described in this chapter out of income. For such items as paths and fuel-bunkers this is quite reasonable. A garage is a more costly item, however, especially if the bricks were not included in the original order for the house bricks. Once again money can be saved by doing this work oneself, and there is the advantage that speed is not as essential as when the house was being built.

There are now a number of pre-cast sectional concrete garages on the market, but size for size these tend to be more expensive than building a brick garage oneself. The prices for concrete garages do not include the prepared base and the drive.

Money can also be saved by building fuel-bunkers, especially as there will probably be enough bricks available from the main building operation.

Again, it is much cheaper to make up one's own sectional shed rather than buy one. It should be possible to construct a reasonable-sized sectional shed for about £20·00.

The average summary for this chapter is as follows:

	£ p
Paint and distemper	15·00
Garage and drive:	
2500 bricks	35·00
Garage doors and frames	18·00
Windows, side door and frame	10·00
Asbestos sheet	13·50
Gravel, 9 yd.	27·00
Cement	16·50
Sand	2·00
Sundries (Purlins, gutter, etc.)	10·00
Paths:	
1 ton cement	11·00
6 yd. gravel	18·00
Front wall:	
2000 bricks	28·00
Half ton cement	5·50
1 yd. sand	2·00
1 yd. gravel	3·00
Gates	12·00
Total	**£226·50**

Let us now take a look at the total of the costs quoted in the previous chapters. The summary above has not been given in two forms as it is assumed that it will be all 'own labour.'

	As built	Using own labour
	£ P	£ P
Chapter 3	110·40	65·40
Chapter 4	224·55	181·65
Chapter 5	470·69	418·44
Chapter 6	725·90	607·90
Chapter 7	372·35	347·35
Chapter 8	175·49	115·49
Chapter 9	552·85	427·85
Scaffolding	50·00	50·00
Total	£2682·23	£2214·08

	As built	Using own labour
	£ P	£ P
Total of Chapters 3–9	2682·23	2214·08
Chapter 10	226·50	226·50
Cost of land	500·00	500·00
Solicitor's fee.	60·00	60·00
Total	3468·73	3000·58
Say	£3500	£3000

The above prices are intended only as a guide. Building materials and labour prices vary greatly in different parts of the country. Even in the interval of going to press some of the prices quoted will be out of date. However, the prices mentioned show how the costs are divided over the complete job, and out of this the materials cost £1890, which is a reasonable average.

The costs have been based on a house of approximately 1000 sq. ft. area, and, again, this tends to be higher than the average.

When the plans of a proposed building go to the council or company granting the mortgage, a value is set on the building from these plans. A detached house with a garage may, for example, be valued at £4000. The actual mortgage granted may be £3600, and this is actual cash available for building. This is enough for the house 'as built' (which means employing direct labour) with a margin of £100. This means that the house is being built without any deposit. If for any reason the cost of building comes above the £3600, the difference must be found by the owner, and it becomes, in effect, a deposit.

In the case of the 'own labour' total, it shows that it should be able to get below the mortgage figure by a comfortable margin. This, in turn, means that the full amount of the mortgage need not be taken up, and this gives a corresponding reduction in the weekly repayments.

As it has been mentioned previously, the more work one is prepared to undertake, the greater the saving. Taken to a logical conclusion, the price of the house could be reduced to the costs of the materials, the cost of the land, and the solicitor's fee! However, not everyone is prepared to go as far as that, and a compromise must be made according to determination and circumstances. One way of cutting costs would be to join forces with another person with the same idea, and build a pair of houses or bungalows. Another method is to start a building group, and there have been many instances of the success of these groups.

Whichever method is adopted, the work involved (which must also include keeping a record of correspondence, bills, invoices, etc.) is hard. The final results, however, more than make it worth while, and if this book gives assistance to anyone engaged on such a project, or inspires anyone to make the effort, it, too, will have been worth while.

Index

H

Hardcore, 56
Hinges, 150

J

Joiner, 28
Jointing, 68
Joists, 72, 76
Joists, preservation, 73, 74

L

Latches, 151
Lavatory basin, 106
Lead, 92, 93
Lighting circuit, 112
Lino tiles, 147
Lintels, 69
Locks, 152

M

Manholes, 122
Mortar, 60

P

Painter, 38
Partitions, 135
Paths, 164
Plasterboards, 140
Plasterer, 36
Plastering, 140
Plugs, 114
Plumber, 31
Power float, 58
Profile boards, 43, 44
Purlins, 88

Q

Quantities of material, 48, 59, 75, 94, 116, 128, 154, 168

R

Rafters, 86
Rafters, pattern, 86
Reinforcing, 70
Ridge board, 86
Ridge tiles, 92
Ring circuit, 114
Rising main, 101

S

Scaffolding, 74, 77, 142
Septic tanks, 127
Services, 53, 56, 100
Setting out, 43
Sheds, 167
Shuttering, 70
Sink, 107
Site, 3
Skirting boards, 150
Skirting, cove base, 147
Soakaways, 127
Soffit, 88
Soil and vent pipe, 120, 125, 126
Soldier arch, 69
Specification, 18, 47, 59, 75, 94, 116, 128, 153, 168
Staircase, 133
Steps, 156, 158, 166
Storage tank, cold water, 104
Storage tank, hot water, 104
Struts, 88
Strutting, 73
Sub-floor, 56

T

Ties, 55
Tiler, 34
Tiles, 90
Tiles, thermoplastic, 146, 147